STECK-VAUGHN

BASIC ESSENTIALS OF MATHEMATICS

BOOK TWO

PERCENT, MEASUREMENT & FORMULAS, EQUATIONS, RATIO & PROPORTION

AUTHOR
James T. Shea

CONSULTANT
Brenda C. Ramsey
ABE/GED Instructor
Memphis City Schools

ACKNOWLEDGMENTS

Senior Math Editor: Karen Lassiter, Ph.D.
Project Coordinator: Joyce Spicer

Project Design and Development:
The Wheetley Company

Cover Design and Illustration:
James Masch

STECK-VAUGHN ADULT EDUCATION ADVISORY COUNCIL

Donna D. Amstutz
Assistant Professor
Northern Illinois University
DeKalb, Illinois

Sharon K. Darling
President, National Center
  for Family Literacy
Louisville, Kentucky

Roberta Pittman
Director, Project C3 Adult Basic Education
Detroit Public Schools
Detroit, Michigan

Elaine Shelton
President, Shelton Associates
Consultant, Competency-Based Adult Education
Austin, Texas

STECK-VAUGHN
ELEMENTARY · SECONDARY · ADULT · LIBRARY

A Harcourt Company

www.steck-vaughn.com

Table of Contents

To the Student

The *Basic Essentials of Mathematics* is a two-book basic math program that teaches whole number, fraction, and decimal skills in Book 1 and teaches percent, measurement, formulas, equations, ratio, and proportion skills in Book 2. Both books have been designed and written for the adult learner who wants to brush up on math skills in a minimum amount of time. These books allow adult learners to act as their own teacher and check their own work. This way, each learner can progress at his or her own rate.

Each page is a separate lesson that deals with a single skill. Many pages contain a blue box at the top where example problems are worked out in a step-by-step procedure. The remainder of the page contains practice problems similar to the ones worked out in the example box at the top of the page. Frequently, the first practice problem is worked out, demonstrating how to arrive at the correct answer.

The Answer Key on page 119 allows students to check their own work. Checking Up pages and Unit Review pages provide frequent review of previously taught skills. These review pages can be used to indicate mastery of the skills taught or the need for additional practice.

To help develop skills in solving problems, three pages of problem-solving strategies appear in each book. These strategies include Choose an Operation, Find a Pattern, Use Estimation, Multi-step Problems, Make a Diagram, and Identify Extra Information. Other lessons develop skills in estimating and rounding numbers.

Each of the two books concludes with a Mastery Test of 100 items. These tests are divided into skill areas. A score chart is included. If a student misses more than two problems in any skill area, he or she should go back and review that skill. When Book 1 is mastered, students may proceed with confidence to Book 2.

Unit 1　Percent

The Meaning of Percent

Stores often advertise items for sale at 10%, 25%, or 20% off the regular price. Percent means **hundredths.** Therefore, "20% off" means that sale items can be purchased at a savings of 20% or 0.20 off the usual price. To find the amount of savings, change the percent to a decimal and multiply.

EXAMPLE: How much will Juan save by buying a $5.00 tie at a 20%-off sale?

Original price:　$5.00	Multiply:　$5.00
Savings rate:　20% = 0.20	✕　0.20
Juan saves $1.00.	$1.00

Remembering that percent means hundredths, solve these problems.

1. Joel Biro bought a suit at a 20%-off sale. The suit was originally priced at $160. How much did he save?

   Answer $32.00

2. Bert bought a hat at a discount of 20%. The hat had been selling for $32. How much did he save?

   Answer $6.40

3. Bert's brother bought a suit at a 20%-off sale. The suit had been marked at $250. How much did he save?

   Answer $50.00

4. Mr. Browne bought six shirts at the same sale. They had been marked $31 each. How much did he save on the six shirts? (First, find how much was saved on one shirt. Then, multiply by 6.)

   Answer _____

5. Anna Browne bought a suit at a 25%-off sale. The suit had been marked $100. How much did she save by buying on sale?

   Answer $25.00

6. Anna also bought a belt which had been marked $25. How much did she save at the 25%-off sale?

   Answer $6.25

7. Angelica bought a dress at this sale. The dress had been marked at $72. How much did she save?

   Answer _____

8. Lila bought a suit at this sale. The suit had been marked to sell for $112. How much did she save by buying at 25% off?

   Answer _____

9. Monty bought a lamp at a discount of 15%. The original price of the lamp was $75. How much did he save on the lamp?

   Answer _____

10. The Rodriguez family bought a new television for $495. It later went on sale for 20% off. How much could they have saved buying the television at the sale price?

    Answer _____

Interchanging Percent and Decimals

To change a *percent to a decimal*, move the decimal point 2 places to the left and drop the % sign.

Remember, when there is no decimal point in a number, it is understood to be at the right of the number. A zero by itself to the *left* of a decimal point is a placeholder (as in 0.6 or 0.02).

EXAMPLES

$64\frac{1}{2}\% = 64.5\% = 0.645$ $7\% = 07\% = 0.07$ $28\% = 0.28$

Change each of these percents to decimals.

1. $20\% = $ *0.20* $15\% = 0.15$ $10\% = 0.10$ $25\% = 0.25$ $50\% = 0.50$ $75\% = 0.75$

2. $30\% = 0.30$ $40\% = 0.40$ $12\% = 0.12$ $18\% = 0.18$ $22\% = 0.22$ $27\% = 0.27$

3. $14.5\% = $ *0.145* $17.6\% = 0.176$ $60.1\% = 0.601$ $33\frac{1}{2}\% = 0.335$ $45\frac{1}{4}\% = $ $35\frac{2}{5}\% = $

4. $90\% = 0.90$ $80\% = 0.80$ $70\% = 0.70$ $65\% = 0.65$ $95\% = 0.95$ $100\% = 1.000$

5. $1\% = $ *0.01* $7\% = 0.07$ $5\% = 0.05$ $9\% = 0.09$ $8\% = 0.08$ $3\% = 0.03$

To change a *decimal to a percent*, move the decimal point 2 places to the right and write a percent symbol. Write zeros as needed.

EXAMPLES

$0.825 = 82.5\%$ or $82\frac{1}{2}\%$ $0.03 = 3\%$ $0.4 = 0.40 = 40\%$

Change the following decimals to percents.

6. $0.10 = $ *10%* $0.20 = 20\%$ $0.30 = 30\%$ $0.40 = 40\%$ $0.50 = 50\%$ $0.60 = 60\%$

7. $0.15 = 15\%$ $0.25 = 25\%$ $0.35 = 35\%$ $0.45 = 45\%$ $0.65 = 65\%$ $0.75 = 75\%$

8. $0.12 = 12\%$ $0.27 = 27\%$ $0.43 = 43\%$ $0.67 = 67\%$ $0.90 = 90\%$ $1.00 = 100\%$

9. $0.01 = $ *1%* $0.05 = 5\%$ $0.09 = 9\%$ $0.07 = 7\%$ $0.03 = 3\%$ $0.08 = 8\%$

10. $0.1 = 1\%$ $0.2 = 2\%$ $0.3 = 3\%$ $0.7 = 7\%$ $0.9 = 9\%$ $0.5 = 5\%$

11. $0.125 = $ *12.5%* $0.375 = 37.5\%$ $0.625 = 62.5\%$ $0.875 = 87.5\%$ $0.085 = 8.5\%$ $0.015 = 1.5\%$
     or *12½%*

6

Changing Fractions to Percent

Any fraction can be written as a percent. To do so, follow these steps.
(1) Change the fraction to its decimal equivalent by dividing the numerator by the denominator.
(2) Move the decimal point two places to the right.
(3) Place the percent sign to the right of the number.

$$\frac{1}{2} \quad 2\overline{)1.00} \quad .50 = 50\%$$

Change $\frac{1}{5}$ to a percent.

$$5\overline{)1.00} \quad 0.20 = 20\%$$

Change $\frac{3}{10}$ to a percent.

$$10\overline{)3.00} \quad 0.30 = 30\%$$

1. $\frac{1}{2} = 50\%$ $\frac{1}{4} = 25\%$ $\frac{3}{4} = 75\%$ $\frac{1}{5} = 20\%$ $\frac{2}{5} = 40\%$

2. $\frac{3}{5} = 60\%$ $\frac{4}{5} = 80\%$ $\frac{3}{10} = 30\%$ $\frac{7}{10} = 70\%$ $\frac{9}{10} = 90\%$

Fractions of halves, fourths, fifths, tenths, and the like, as you have just seen, produce **even** two-place decimals. Other fractions (like eighths, twelfths, and sixteenths) can be changed to decimals if the division is carried out more than two places.

Change $\frac{1}{8}$ to a percent.

$$8\overline{)1.000} \quad 0.125 = 12.5\% \text{ or } 12\tfrac{1}{2}\%$$

Change $\frac{3}{16}$ to a percent.

$$16\overline{)3.0000} \quad 0.1875 = 18.75\% \text{ or } 18\tfrac{3}{4}\%$$

Change the following fractions to percents.

3. $\frac{3}{8} = 37.5\%$ $\frac{5}{8} = 62.5\%$ $\frac{7}{8} = 87.5\%$ $\frac{1}{32} = 0.32\%$ $\frac{5}{32} =$

4. $\frac{1}{16} =$ $\frac{3}{16} =$ $\frac{5}{16} =$ $\frac{7}{16} =$ $\frac{11}{16} =$

When some fractions are changed to decimals, there are always remainders, no matter how many places we carry out the division. In such cases, three, or sometimes four, places is far enough to carry the division. Then the remainder is expressed as a fraction.

$$9\overline{)1.0000} \quad .1111\tfrac{1}{9} \quad 11\tfrac{1}{9}\%$$

Solve the following problems.

5. **Change $\frac{1}{3}$ to a percent:**

$$3\overline{)1.0000} \quad 0.3333\tfrac{1}{3} = 33\tfrac{1}{3}\%$$

Change $\frac{1}{9}$ to a percent:

Change $\frac{2}{3}$ to a percent:

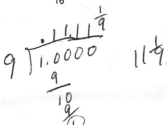

$$3\overline{)2.000} \quad .666\tfrac{2}{3} \quad 66\tfrac{2}{3}\%$$

7

Fractions, Decimals, and Percents

All percents can be changed to fractions. Many percents can be changed to very simple fractions. In many cases it is easier to use a simple fraction than to use a decimal. To change a percent to a fraction, place the percent over 100 and drop the % sign. Simplify.

Fill in the missing equivalent forms.

ϕ 3/6

Do Your Work Here

$$50\% = 0.50 = \frac{50}{100} = \frac{1}{2}$$

	FRACTION	DECIMAL	PERCENT
1.	$\frac{1}{2}$	*0.50*	50%
2.	$\frac{1}{4}$	0.25	25 %
3.	$\frac{3}{4}$	0.75	75%
4.	$\frac{2}{5}$	0.40	40%
5.	$\frac{3}{5}$	0.60	60%
6.	$\frac{1}{10}$	0.10	10%
7.	$\frac{7}{10}$	0.70	70%
8.	$\frac{1}{4}$	0.25	25%
9.	$\frac{3}{10}$	0.30	30%
10.	$\frac{625}{1000}$	0.625	62.5%
11.	$\frac{9}{20}$	0.45	45%
12.	$\frac{3}{20}$	0.15	15%
13.	$\frac{72}{100}$	0.72	72%
14.	$\frac{7}{50}$	0.14	14%
15.	$\frac{1}{25}$	0.04	4%
16.	$\frac{5}{100}$	0.05	5%
17.	$\frac{15}{100}$	0.15	15%
18.	$\frac{1}{5}$	0.20	20%
19.	$\frac{1}{8}$	0.125	12.5%
20.	$\frac{375}{1000}$	0.375	37.5%
21.	$\frac{33}{100}$	0.33	$33\frac{1}{3}$%
22.	$\frac{2}{3}$	0.66	$66\frac{1}{6}$%
23.	$\frac{4}{5}$	0.80	80%
24.	$\frac{1}{12}$	0.08	8%
25.	$\frac{625}{1000}$	0.62.5	$62\frac{1}{2}$%
26.	$\frac{875}{1000}$	0.87.5	$87\frac{1}{2}$%
27.	$\frac{4}{5}$	0.80	80%
28.	$\frac{9}{10}$	0.90	90%
29.	$\frac{7}{10}$	0.70	70%

8

Solve.

1. The average household spends $0.35 of every dollar earned on housing costs. What percent is this amount?

   35 %

2. Three out of four students passed the exam. What percent of the students passed the exam?

   75 %

3. Real estate taxes in the city increased by 7.25%. What decimal number would you use to figure the new real estate taxes?

   0.725

4. In the last election, 36% of the people voted "Yes" on Proposition 6. What fraction of the people is this?

   36/100

5. The bank used 0.0975 to figure the interest charge on the McCall's mortgage. What percent is this?

   9.75 %

6. On the average, 17 out of every 100 viewing minutes is taken up by commercials. What is this average as a percent?

   17 %

7. Sales tax in the city is $5\frac{1}{2}$%. Write the decimal number used to figure the sales tax on a purchase.

   5.5   .055

8. Home improvement costs have increased by $\frac{2}{5}$ in the last few years. By what percent have home improvement costs increased?

   40 %

9. Only one percent of the population lives in solar homes. What decimal would you use to find how many people live in solar homes?

   0.01

10. One brand of ham is 99.5% fat free. Write this percent as a decimal.

    99.5

11. Four out of five people chew sugarless gum. What percent chew sugarless gum?

    80 %

12. The interest rate on some home mortgages is 12 cents for every dollar. What is this amount as a percent?

    12 %

9

Percents Larger than 100%

Large percents are used just like the percents we have been using. To change a percent to a decimal, move the decimal point 2 places to the left. Drop the percent sign.

EXAMPLES

$$200\% = 2.00 = 2 \qquad 375\% = 3.75 \qquad 105\frac{3}{10}\% = 105.3\% = 1.053$$

Change each percent to a decimal.

1. $300\% =$ _3_ $400\% =$ _4_ $600\% =$ _6_ $800\% =$ _8_ $1{,}000\% =$ _10_

2. $150\% =$ _1.50_ $125\% =$ _1.25_ $175\% =$ _1.75_ $199\% =$ _1.99_ $500\% =$ _5_

3. $225\% =$ _2.25_ $250\% =$ _2.50_ $112\frac{1}{2}\% =$ _1.125_ $137\frac{1}{2}\% =$ _1.375_ $162\frac{1}{2}\% =$ _1.625_

4. $310\% =$ _3.10_ $337\frac{1}{2}\% =$ _3.375_ $387\frac{3}{4}\% =$ _3.8775_ $133\frac{1}{5}\% =$ _1.3326_ $166\frac{2}{5}\% =$ _1.66_

5. $101\% =$ _1.01_ $160\% =$ _1.60_ $105\% =$ _1.05_ $120\% =$ _1.20_ $202\% =$ _2.02_

6. $205\% =$ _2.05_ $999.9\% =$ _9.099_ $450\% =$ _4.50_ $807\% =$ _807_ $320\% =$ _3.20_

7. $609\% =$ _6.09_ $725\frac{1}{2}\% =$ _7.255_ $1{,}100\% =$ _11.00_ $101\frac{3}{10}\% =$ _1.013_ $175\frac{1}{4}\% =$ _1.7525_

Solve.

Do Your Work Here

8. A clothing store has a 210% markup on the price of their clothing. Write the decimal used to figure the price of their clothing.   _2.10_

9. Larry's test scores have improved 152%. Write this percent as a decimal.   _1.52_

10. The profits for Fun Times, Inc. have increased 333% since the company started. Write this amount as a decimal.   _3.33_

11. Doreen's savings have increased by 135%. What decimal would you use to figure the amount of money in Doreen's account?   _1.35_

12. An automobile company's sales have increased 116% in two years. Write this increase as a decimal.   _1.16_

13. A jeweler uses a 300% markup on necklaces. Write this amount as a decimal.   _3_

Percents Smaller than 1%

Small percents can be used just like the large percents we have been using. Proceed as you did with large percents. To solve problems using percents smaller than 1%, change the fraction to a decimal.

(handwritten: 1% 0.19 1/10   15%)
(handwritten: .91   .001%   .15)

EXAMPLES

$0.1\% = 0.001$ $\frac{3}{4}\% = 0.75\% = 0.0075$

Change each percent to a decimal.

1. $0.5\% = \underline{0.005}$ $0.33\% = \underline{0.0033}$ $0.25\% = \underline{0.0025}$ $0.6\% = \underline{0.006}$

2. $0.4\% = \underline{0.004}$ $0.41\% = \underline{0.0041}$ $0.05\% = \underline{0.0005}$ $0.75\% = \underline{0.0075}$

3. $\frac{1}{4}\% = \underline{0.25\%}$ $\frac{2}{5}\% = \underline{0.0040}$ $\frac{3}{5}\% = \underline{0.0060}$ $\frac{4}{5}\% = \underline{0.0080}$ $\frac{1}{8}\% = \underline{0.00125}$
   $= \underline{0.0025}$

4. $\frac{3}{8}\% = \underline{0.00375}$ $\frac{5}{8}\% = \underline{0.00625}$ $\frac{7}{8}\% = \underline{0.00875}$ $\frac{1}{10}\% = \underline{0.001}$ $\frac{3}{10}\% = \underline{0.003}$

5. $\frac{7}{10}\% = \underline{0.007}$ $\frac{9}{10}\% = \underline{0.009}$ $\frac{5}{16}\% = \underline{0.003125}$ $\frac{7}{16}\% = \underline{}$ $\frac{1}{20}\% = \underline{0.0005}$

Solve.

Do Your Work Here

6. Laurel's weight decreased by 0.9% since she last weighed herself. Write the decimal you would use to find Laurel's new weight.

   0.009

7. The measurements were off by 0.64%. Write the number you would use to show the error.

   0.0064

8. Joanne's test scores improved by 0.55%. What number would you use to find Joanne's test scores?

   0.0055

9. Which is greater, 0.45% or $\frac{5}{16}\%$? Write each as a decimal and compare.

   $\frac{5}{16}\%$

10. A soap advertises that it is $99\frac{44}{100}\%$ pure. If this is true, what percent represents the impurities?

    0.001%
    $\frac{999}{1000}$

11. You have heard the expression, "Nine hundred ninety-nine times out of a thousand." Express this as a fraction, as a percent, and as a decimal.

    0.0099%
    0.999%

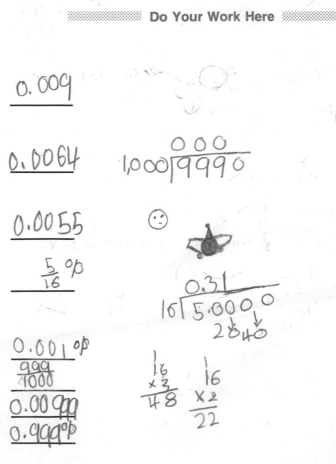

11

Finding a Percent of a Number

To find a percent of a number, write a percent sentence. Every percent sentence consists of three numbers: the rate, the whole, and the part.

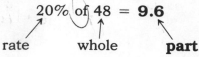

$$20\% \text{ of } 48 = \mathbf{9.6}$$

rate   whole   **part**

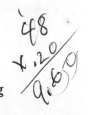

If the *part* is missing in a percent problem, solve by first changing the *rate* to a decimal. Then multiply the rate by the *whole*.

Remember, "of" means multiply.

Find: 25% of 84

$25\% \times 84 = ?$	$\begin{array}{r} 8\,4 \\ \times\,0.2\,5 \\ \hline 4\,2\,0 \\ 1\,6\,8\,0 \\ \hline 2\,1.0\,0 \end{array}$
$25\% = 0.25$	
$0.25 \times 84 = 21$	

Find: 105% of 280

$105\% \times 280 = ?$	$\begin{array}{r} 2\,8\,0 \\ \times\,1.0\,5 \\ \hline 1\,4\,0\,0 \\ 0\,0\,0\,0 \\ 2\,8\,0\,0\,0 \\ \hline 2\,9\,4.0\,0 \end{array}$
$105\% = 1.05$	
$1.05 \times 280 = 294$	

Change each percent to a decimal. Solve.

1. 50% of 90   300% of 60   50% of 60   75% of 120   35% of 60
   $0.50 \times 90 = 45$ 180   30   90   480

2. 80% of 120   75% of 80   10% of 180   20% of 60   25% of 45
   96   60   18   12   11.25

3. 40% of 75   90% of 200   25% of 80   21% of 200   60% of 160
   30   180

4. 250% of 100   99% of 55   25% of 400   40% of 150   14% of 500
   25000   54.45   100   42.   70

5. 8% of 900   35% of 700   90% of 640   75% of 400   30% of 120

6. 36% of 125   5% of 860   15% of 450   45% of 40   10% of 90

7. 15% of 260   25% of 600   125% of 80   225% of 12   5% of 843
   39   150   100   27   42.15

Problems Using Percents

Use percent to solve these word problems.

1. A real estate agent sold our house to Ira Brown for $65,000. For her service in making the sale, the agent charged 5% of the sale price. How much did we have to pay the agent?

Answer _____325_____

2. Thomas says that he spends 25% of his salary for rent. If his annual salary is $24,000, how much total rent does he pay each year?

Answer _____5 800_____

3. Thomas said that 30% of his earnings was spent for food. How much money does he spend in one year for food?

Answer _____7 200_____

4. He also said that he spent 15% of his salary for clothing for the family. How much did he spend last year for clothing?

Answer _____3600_____

5. For insurance and savings Thomas said that he had set aside 20% of his salary. How much did he set aside last year?

Answer _____4800_____

6. Thomas also said that doctors, hospitals, medicine, and medical costs took 8% of his salary. How much was spent for these?

Answer _____1920_____

7. The Shamrock Motor Sales Co. sells 40% of all the cars sold in the town of Polanco. Last year 500 cars were sold in the town. How many were sold by the Shamrock Co.?

Answer _____200 cars_____

8. A bushel of potatoes weighs 60 pounds. It is said that 50% of this weight is made up by the water in potatoes. How many pounds of water are there in a bushel of potatoes?

Answer _____30 pounds lbs._____

9. A gallon of milk weighs (on the average) 8.6 pounds. Of this amount 75% is water. In the average gallon of milk, how much does the water weigh?

Answer _____6.2_____

lbs.
$$48.6$$
$$.75$$
$$430$$
$$602$$
$$6.450$$
6.45

10. Butterfat makes up 4% of the milk. How many pounds of butterfat are there in the average gallon of milk?

Answer _____3.44 lbs_____ 3/17

13

Finding a Percent of a Number
Using Fractions

Another way to find a percent of a number is to change the *percent* to a fraction. You must also change the *whole* to a fraction. Then multiply. Often it is easier to multiply by a fraction than by a decimal.

Remember, to change a percent to a fraction:
- Write the percent as a decimal.
- Drop the decimal point and place the number over 100.
- Simplify.

Find: 20% of 50

$$20\% \times 50 = ?$$

$$20\% = 0.20 = \frac{20}{100} = \frac{1}{5}$$

$$\frac{1}{\cancel{5}} \times \frac{\cancel{50}^{10}}{1} = \frac{10}{1} = 10$$

Find: 75% of $1\frac{3}{4}$

$$75\% \times 1\frac{3}{4} = ?$$

$$75\% = 0.75 = \frac{75}{100} = \frac{3}{4}$$

$$\frac{3}{4} \times \frac{7}{4} = \frac{21}{16} = 1\frac{5}{16}$$

Change each percent to a fraction. Solve.

1. 25% of 16
    $$25\% = 0.25 = \frac{25}{100} = \frac{1}{4}$$
    $$\frac{1}{\cancel{4}} \times \frac{\cancel{16}^{4}}{1} = \frac{4}{1} = 4$$

    50% of $1\frac{1}{3}$ $= \frac{2}{3}$

    5% of 60 6

2. 85% of 40 $\frac{34}{1}$ 34

    40% of $3\frac{1}{3}$ $= 1\frac{1}{3}$

    10% of 90 $\frac{9}{1}$

3. 60% of 55 $\frac{33}{1}$ 33

    80% of $3\frac{3}{4}$ $\frac{3}{1}$ 3

    15% of 80 $\frac{12}{100}$

4. 6% of 20 $1\frac{1}{5}$

    70% of $6\frac{2}{3}$ $4\frac{2}{3}$

    30% of 50 15

5. 75% of $4\frac{1}{2}$

    50% of 16

    20% of 25

6. 90% of $2\frac{1}{2}$

    35% of 40

    8% of 35

7. 18% of 50

    45% of 80

    85% of $3\frac{1}{2}$

14

The Meaning of Simple Interest

Interest is a fee or rental charge paid for the use of money. The bank uses a customer's savings money for loans to its other customers and, in turn, pays the saver for the use of his or her money. Interest can be considered as rent for the use of money.

Simple interest is always figured on the basis of a year. The percent of interest is called the **rate.** The money involved is called the **principal.** The number of years involved is called the **time.**

Interest is figured using the formula
$I = prt$ (interest = principal × rate × time).

EXAMPLE: Patricia has a savings account at a bank in her town. Her savings account amounts to $100. The bank pays 5% on all savings accounts. How much interest will she get at the end of one year?

p (principal) = *$100*	$I = prt$
r (rate) = *5%* = *0.05*	I (interest) = $p \times r \times t$
t (number of years) = *1*	$I = $100 \times 0.05 \times 1 = 5.00

Using the same formula, find the interest on each of the following sums.

1. $500 at 2% for 1 year	**2.** $225 at 4% for 1 year	**3.** $650 at 4% for 1 year
$5.10 10	$234 9	$676 26
4. $200 at $3\frac{1}{2}$% for 1 year	**5.** $800 at $2\frac{1}{2}$% for 1 year	**6.** $500 at 6% for 1 year
7.00	20	30
7. $1,000 at 4% for 1 year	**8.** $700 at $3\frac{1}{2}$% for 1 year	**9.** $300 at $2\frac{1}{2}$% for 1 year

Solve the problems below.

10. Agnes has had $350 in a savings account for a year. She gets $5\frac{1}{4}$% yearly interest. How much interest will she get? (Round your answer to the nearest cent.)

11. Dan's bank pays 5% annual interest. How much interest will he have on his savings account of $360 after a year?

$378

12. Jerry Jennings borrowed $750 from a bank for 1 year. The bank charged 8% annual interest on the loan. How much interest will he have to pay?

60

More on Simple Interest

Interest is paid on the basis of a year's time. But not all accounts are in the bank for an exact number of years. Neither are all loans made for whole years. Some loans are made for less than a year. Other loans are made for more than a year. Study the examples below to see how to figure interest on such loans.

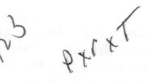

EXAMPLE: Find the interest on $300 at 4% for 3 months.

$$I = prt$$
$$p = \$300$$
$$r = 4\% = 0.04$$
$$t = 3 \text{ mo} = \frac{3}{12} \text{ yr} = \frac{1}{4}$$
$$I = \$300 \times 0.04 \times \frac{1}{4} = \$3.00$$

EXAMPLE: Find the interest on $5,000 at 13% for $4\frac{1}{2}$ years.

$$I = prt$$
$$p = \$5,000$$
$$r = 13\% = 0.13$$
$$t = 4\frac{1}{2} = \frac{9}{2}$$
$$I = \$5,000 \times 0.13 \times \frac{9}{2} = \$2,925.00$$

Find interest on each of the following. Be sure to change months to years.

1. $300 at 2% for 4 months $2.00	**2.** $600 at 4% for 10 months $20.00	**3.** $600 at $2\frac{1}{2}$% for 4 months $50.00
4. $600 at 4% for 7 months $14.00	**5.** $1,000 at 6% for 6 months $30.00	**6.** $1,200 at 2% for 9 months $18.00
7. $400 at 2% for $2\frac{1}{2}$ years	**8.** $450 at $12\frac{1}{2}$% for 2 years	**9.** $600 at 3% for 2 years and 3 months
10. $1,000 at 14% for 3 years and 6 months so hello	**11.** $960 at 4% for 2 years and 4 months	**12.** $1,200 at 11% for 3 years and 10 months

16

The Meaning of Compound Interest

In most savings accounts, interest earned during a period is added to the account to form a new principal. The interest for the next period is earned on this new principal. This process of paying interest on interest is called **compound interest.** All savings banks and other savings organizations pay compound interest. In compound interest, interest is paid on interest. Compound interest is paid annually (every year), semiannually (every six months), quarterly (every three months), or daily, depending upon the regulations of the savings institution.

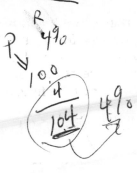

Study the example below before solving the problems on this page.

EXAMPLE: Thomas has $200 in a savings account paying 4% interest compounded semiannually. How much will he have in the bank at the end of the year?

$p = \$200$
$r = 4\% = 0.04$
$t = \frac{1}{2}$
$I = prt = \$200 \times 0.04 \times \frac{1}{2} = \$4.00$
$\$200 + \$4 = \$204.00$
(amount in the bank at the end of $\frac{1}{2}$ year)

$p = \$204$ (at the end of $\frac{1}{2}$ year)
$r = 4\% = 0.04$
$t = \frac{1}{2}$
$I = prt = \$204 \times 0.04 \times \frac{1}{2} = \$4.08$
$\$204.00 + \$4.08 = \$208.08$
(amount in the bank at the end of 1 year)

Solve.

1. Julia has $300 in the savings bank and is paid 4% semiannually. How much will she have at the end of a year?

   Answer __$306__

2. Marguerite's bank pays $4\frac{1}{2}\%$ interest on a semiannual basis. How much will she have in the bank at the end of the first year if she has $400 at the start of the year? (For each step, round your answer to the nearest cent.)

   418.20

   Answer __$405__

3. Franco has $500 in a savings account and is paid 5% semiannually. How much will he have at the beginning of the second year? (For each step, round your answer to the nearest cent.)

   Answer __$538.49__

4. Maria has $300 in her savings account and is paid 5% compounded semiannually. How much will she have in her account at the end of the second year? (For each step, round your answer to the nearest cent.)

   Answer __331.15__

26.27

17

Percent of Increase

To find the current value or amount, first multiply the original amount by the percent of increase. Then add.

EXAMPLE: The rent for a 1-bedroom apartment was $275 per month. This year the rent went up 4%. How much is the apartment renting for this year?

original amount	$ 2 7 5		original amount	$ 2 7 5
percent of increase	× 0.0 4		amount of increase	+ 1 1
amount of increase	*$1 1.0 0*		current rent	*$2 8 6*

The apartment is renting for $286 this year.

Solve.

1. The federal census shows that one of the nation's fastest-growing cities has increased 92% from its previous population of 256,000. What is the new population figure?

   Answer _____491,520.00_____

2. Arthur Hodgson is a salesman and works on commission. Last year he sold $245,000 in merchandise. This year he did 20% more business. How much did he sell this year?

   Answer _____7,350.00_____

3. Cassy Burleson was making $96.40 a day and was given a 10% increase. How much does she get with the increase?

   Answer _____9,736.40_____

4. This year Mr. Ogilvie, a salesman, increased his sales by 20% over last year. He sold $22,500 worth of merchandise last year. How much did he sell this year?

   Answer _____27,000.00_____

5. The census for a year ago showed that Podunk had a population of 1,640. This year's census shows an increase of $12\frac{1}{2}\%$. What is the population figure for this year?

   Answer _____

6. A farmer averaged 40 bushels of potatoes to the acre on unfertilized ground. This year he has used a fertilizer which is guaranteed to double the production rate. How many bushels to the acre should he raise this year? (Hint: Double means 200 percent.)

   Answer _____

7. Raul Garcia did 250% more business during the past month than he did the month before, when he did $2,650 worth of business. How much did he do last month?

   Answer _____

8. Karen grows tomatoes. This year she produced 300 pounds. She plans to increase production by 20%. How many pounds will she produce after the increase?

   Answer _____

Percent of Decrease

To find the current value or amount, first multiply the original amount by the percent of decrease. Then subtract.

EXAMPLE: Last year's sales figure at Klaus Clothiers was $850,000. This year the total sales figure decreased by 3%. What was this year's sales figure?

original figure	**$850,000**	original figure	**$850,000**
percent of decrease	× **0.03**	amount of decrease	− **25,500**
amount of decrease	*$25,500.00*	new sales figure	*$824,500*

This year's sales figure at Klaus Clothiers was $824,500.

Solve the problems which follow.

---

1. Bernard bought a used car last year for $3,500. It is now worth 25% less (decrease). How much is it worth now?

Answer _____

2. Many stores sell cooked hams. A cook will tell you that a ham shrinks (decreases) while cooking. If a 16-pound ham loses $12\frac{1}{2}\%$ while cooking, how much will it weigh after it is cooked?

Answer _____

---

3. Strawberries decrease, too, when cooked. If they decrease 15%, how much would remain if you started with 16 quarts?

Answer _____

4. Last month our utility bill was $98. This month shows a decrease of $12\frac{1}{2}\%$. How much is this month's bill?

Answer _____

---

5. The storekeeper decreased by 20% the price of a coat which had been marked $85. What was the sale price?

Answer _____

6. Last year we had a total rainfall of 9.6 inches. This year there has been a decrease of 25%. How much is this year's fall of rain?

Answer _____

---

7. The last census shows that the town of Millers has decreased 10% in population. If the previous census figure was 2,400, what is the figure now?

Answer _____

8. Yolanda Gonzales did business last year to the amount of $24,500. This year business has fallen off 15%. To how much will this year's business amount?

Answer _____

Finding Commissions and Net Amount

Fees or commissions are often expressed as percentages. The amount of a commission can be found by multiplying the gross amount by the commission rate. To find the net proceeds, subtract the commission from the gross amount.

> gross amount = total collected
> commission = fee paid for a service
> commission rate = percent of gross amount paid for a service
> net = amount remaining after commission is paid

EXAMPLE: A salesperson sold $400 worth of lumber. How much will the owner of the lumberyard get after paying a 15% commission to the salesperson?

> Amount collected = $400
> Commission rate = 15% = 0.15
> Commission = $400 × 0.15 = $60
> Net proceeds = $400 − $60 = $340

Solve the following problems.

Do Your Work Here

1. Ms. Lee collects accounts for several firms, for which she is paid 7%. Last month she collected $12,400. How much did the firms have after paying Ms. Lee?   _$11,532_

$$
\begin{array}{r}
\$12,400 \\
\times \quad 0.07 \\
\hline
\$868.00
\end{array}
\qquad
\begin{array}{r}
\$12,400 \\
- \quad 868 \\
\hline
\$11,532
\end{array}
$$

2. Walter sold a house for $87,600. He paid the real estate agent 5% for the sale. How much did Walter net (receive) from the sale?   _____

3. A cosmetics salesman sold a total of $5,260 in one month. His commission was $12\frac{1}{2}\%$. How much did he make?   _____

4. A real estate agent sold a vacant lot for us at $4,220. He charged 10%. How much did we receive net from the sale?   _____

5. An appliance sales clerk sold 2 stoves at $345 each and 2 radios at $65 each. She was paid a commission of 10%. How much did the store net?   _____

6. A rancher shipped 40,000 pounds of cattle to a commission house. The cattle were sold for $32,000, and the commission house charged 6%. How much did the rancher net?   _____

7. An auctioneer sold $1,240 worth of furniture for Miss Coffey. He charged a fee of 5% as his commission. How much did Miss Coffey receive from the sale?   _____

20

Finding a Number When a Percent of It Is Known

To find the *whole* in a percent problem, write a percent sentence.
Change the *rate* to a decimal. Divide the *part* by the decimal.

20% of **48** = 9.6

rate   **whole**   part

(handwritten: 12% × n = 18, n = 18 ÷ .12)

12% of what number is 18?

> 12% × ? = 18
> 12% = 0.12
> ? = 18 ÷ 0.12
> 150 = 18 ÷ 0.12

30% of what number is 180?

> 30% × ? = 180
> 30% = 0.30
> ? = 180 ÷ 0.30
> 600 = 180 ÷ 0.30

Change each percent to a decimal. Solve.

1. 25% of what number is 17?
   > 25% = 0.25
   > ? = 17 ÷ 0.25
   > 68 = 17 ÷ 0.25

   32% of what number is 40? *115*

   80% of what number is 32? *40*

2. 80% of what number is 64? *80*

   135% of what number is 270? *2,000*

   5% of what number is 2,850? *5,500*

3. 60% of what number is 33? *55*

   45% of what number is 90? *200*

   40% of what number is 5,200? *13,000*

4. 10% of what number is 73?

   75% of what number is 120?

   90% of what number is 36?

5. 40% of what number is 80?

   80% of what number is 6?

   35% of what number is 7?

6. 12% of what number is 3?

   5% of what number is 15?

   20% of what number is 7?

7. 15% of what number is 30?

   30% of what number is 27?

   60% of what number is 540?

8. 50% of what number is 40?

   10% of what number is 54?

   12% of what number is 6?

3/29

Finding a Number When a Fraction of It Is Known

To find the *whole* using fractions, divide the known quantity by the fraction which describes it. Study the example below.

EXAMPLE: Mr. Franco paid $42 for $\frac{3}{4}$ ton of hay. What was the price of the hay per ton?

$\frac{3}{4} \times ? = \$42$

$? = 42 \div \frac{3}{4}$

Remember, to divide by a fraction, multiply by the reciprocal (invert and multiply).

$\frac{\overset{14}{\cancel{42}}}{1} \times \frac{4}{\cancel{3}} = 56$

The price per ton was $56.

Solve.

1. Two thirds of the families in Smithville own their own homes. The total number of home-owning families there is 480. How many families live in Smithville?

   Answer _____720_____

2. Mr. Thomas spent $32,000 for labor in the construction of his house. The cost of the labor represented $\frac{4}{5}$ of the total cost. What was the total cost of the house?

   Answer _____40,000_____

3. Carmen has saved $16, which is only $\frac{4}{5}$ as much as her brother Tom has saved. How much has Tom saved?

   Answer _____20_____

4. Dale Harvey has two fields in grain. One field is 50 acres and is $\frac{2}{5}$ as large as the other field. How large is the other field?

   Answer _____125_____

5. Ms. Griffin spent $45,000 for labor in the construction of her new home. This is $\frac{3}{5}$ of the total cost of the house. What was the total cost?

   Answer _____

6. Mr. Tompkins planted $\frac{2}{3}$ of his farm in potatoes. This represented 240 acres. His entire farm comprises how many acres?

   Answer _____

7. Sal Bobinsky said that his expenses, totaling $12,000 for the year, amounted to $\frac{1}{3}$ of his income. How much was his yearly income?

   Answer _____

8. If $\frac{3}{4}$ of the weight of a watermelon is water, and if, in the melon which Takashi bought, the part which is water weighs 12 pounds, how much does the whole watermelon weigh?

   Answer _____

22

Finding the Original Price

If you know the discounted price of an item and the percentage of the discount, you can find the original price by using the fraction or decimal method.

EXAMPLE: A store advertised a radio on sale for $24. This was 25% off the regular price. Find the regular price of the radio.

Original price = 100%

Sale price = $100\% - 25\% = 75\% = 0.75 = \frac{3}{4}$

Fraction Method: Original price = $\$24 \div \frac{3}{4} = \frac{24}{1} \times \frac{4}{3} = \frac{8 \times 4}{1 \times 1} = \frac{32}{1} = \$32$

Decimal Method: Original price = $\$24 \div 0.75 = \$32.00$

Solve.

1. Ms. Franklin bought a car for $10,540. This was 15% off the original price. What was the original price?

Answer $12,400$

2. Katie bought a blouse for $24. This was 20% off the original price. What was the original price?

Answer $\$120$ $\$30.00$

3. Felipa bought a pair of sandals on sale at $33\frac{1}{3}\%$ off. She paid $12 for the shoes. What was the original price?

Answer 1 18

4. Mrs. Harper bought a painting for $60, which was 75% of its former price. What was the former price? (Do not subtract the 75% from 100%. Why?)

Answer 300 80

5. By paying cash, Mrs. Perkins got a 5% discount on furniture. She paid $950 for the furniture. What would it have cost without the discount?

Answer $1,000$

6. The school bought books for the library and paid $970 for them, having been given a 3% discount. How much would the books have cost without the discount?

Answer $1,000$

7. The department store had a 15%-off sale on all merchandise. They took in $8,500 on the first day of the sale. If they had sold the same merchandise at the original price, how much would have been taken in?

Answer $10,000$

8. The grocer settled some old accounts for 16% less than the original amounts. On this basis she collected $1,680. What was the total of the original accounts?

Answer $2,000$

Using Fractions or Decimals to Find the Original Amount

Solve.

1. Elena earned $400 last week. This is 80% of the amount she earned the week before. What amount did she earn the previous week?

   Answer $2,000

2. Last month, Juan made a $540 mortgage payment. This is 25% of his monthly income. What is Juan's monthly income?

   Answer $720

3. If Mr. Miller sold 75% of his farm, and the number of acres sold was 150, how many acres were there in the farm before selling?

   Answer 606 acres

4. Figures reveal that 20% of the total population of a community is enrolled in school. In Smithville there are 480 pupils enrolled. What is the total population of Smithville?

   Answer $00 people

5. Pat Bain paid $5,775 for a car. She said that she was allowed $12\frac{1}{2}$% off because the car had been used as a demonstrator. What was the original price of the car? (If she was allowed $12\frac{1}{2}$% off, what percent did she pay?)

   Answer

6. The distance from New York to San Francisco is 5,200 miles by way of the Panama Canal. This is 40% of the distance by way of the Strait of Magellan. How far is it by way of the Strait of Magellan?

   13,000

   Answer

7. Elise has $360 in the savings bank now. This is 20% less than she had a year ago. How much did she have a year ago?

   Answer 1800

8. The rug in the living room contains 12 square yards. It covers 75% of the floor. How many square yards of floor surface are there?

   Answer 48 sy

9. The population of this town is now 1,500, which is an increase of 20% over the previous figure. What was the previous figure? (The increase of 20% makes the present figure 120% of the previous figure. Divide by 1.20.)

   Answer

10. Joe now weighs 180 pounds. This is a decrease of 10% in his weight of a year ago. How much did he weigh then? (10% decrease of 100% leaves 90% for his weight now.)

    Answer

Finding What Percent One Number Is of Another

To find the *rate* in a percent problem, write a percent sentence. Divide the *part* by the *whole*. Then write the decimal answer as a percent.

$$20\% \text{ of } 48 = 9.6$$

rate   whole   part

What percent of 75 is 60?

$? \% \times 75 = 60$
$? = 60 \div 75$
$0.8 = 60 \div 75$
$0.8 = 80\%$

What percent of 160 is 4?

$? \% \times 160 = 4$
$? = 4 \div 160$
$0.025 = 4 \div 160$
$0.025 = 2.5\%$

Find the rate.

1. What percent of 40 is 8?

   $? = 8 \div 40$
   $0.2 = 8 \div 40$
   $0.2 = 20\%$

   What percent of 25 is 15?

   60%

   What percent of 400 is 30?

   50%   7.5

2. What percent of 48 is 7.2?

   15.00%

   What percent of 23 is 6.9?

   $\% \times 36 = 6.9$
   $= 6.9 \div 36$

   What percent of 250 is 10?

   4%

3. What percent of 180 is 81?

   45.0%

   What percent of $35 is $5.25?

   15%

   What percent of 480 is 408?

4. What percent of 400 is 268?

   65.0%

   What percent of 140 is 112?

   80 0%

   What percent of 50 is 40?

   80%

5. What percent of 150 is 16?

   10.%%

   What percent of 40 is 15?

   37.5%

   What percent of 32 is 24?

   75%

6. What percent of 50 is 6?

   What percent of 45 is 9?

   What percent of 75 is 30?

7. What percent of 56 is 7?

   What percent of 120 is 12?

   What percent of 40 is 35?

8. What percent of 8 is 2?

   What percent of 12 is 8?

   What percent of 20 is 7?

Finding What Percent One Number Is of Another

Solve these word problems.

1. Dan bought a $30 radio for $24 at a sale. What percent discount did he get?

$30 − $24 = $6 discount
What percent of $30 is $6?
   ? = 6 ÷ 30
0.20 = 6 ÷ 30
0.20 = 20%

Answer ___20% discount___

2. Juanita had 15 raffle tickets to sell. She sold 6 tickets. What percent of the tickets did she sell?

Answer _____

3. Arthur bought a used car last year for $2,400. He traded it in on a new car this year and was allowed only $1,800 for it. What percent of its previous value was his old car worth?

Answer _____

4. Madge bought a scarf marked $3 for $2.25. What percent did she save?

Answer _____

5. By paying cash, Mr. Terry bought a television for $360. Otherwise it would have cost him $400. What percent did he save?

Answer _____

6. Thomas paid $7.50 for a football which was originally priced at $10. What percent did Thomas save?

Answer _____

7. Virginia bought a $15 shirt on sale for $10. What percent did she save?

Answer _____

8. Dorothy paid $35 for a dress which had been marked $40. What percent did she save?

Answer _____

Checking Up

Fill the blanks in the items below.

1. $33\frac{1}{3}$% of 99 is _329.67_.

2. $66\frac{2}{3}$% of 48 is _____.

3. _25_% of 64 is 16.

4. $33\frac{1}{3}$% of _____ is 36.

5. $12\frac{1}{2}$% of 8 is _1.00_.

6. 150% of _5 10_ is 15.

7. _50 200_% of 14 is 28.

8. 20% of _80_ is 16.

9. 140% of 25 is _3.50_ Decimal.

10. 17% of 50 is _8.50_.

11. _80_% of 5 is 4.

12. 150% of 9 is _13.50_.

13. $\frac{1}{2}$% of 100 is _50_ Decimal.

14. 101% of 500 is _505.00_.

15. 200% of _5_ is 10.

16. _2_% of 700 is 14. Decimal

17. $\frac{1}{2}$% of 400 is _2.00_.

18. 60% of 15 is _6 9_.

19. 25% of _160_ is 40.

20. 50% of $21.50 is _10.75_.

21. 25% of 36 is _9.00_.

22. _9_% of 100 is 9.

23. $62\frac{1}{2}$% of _____ is 15.

24. 2% of 2,000 is _40_.

25. 105% of 20 is _24_.

26. _16_% of 80 is 20.

27. 1.5% of 44 is _6.6_.

28. _200_% of 101 is 202.

29. 0.5% of 2,000 is _10_.

30. _20_% of 85 is 17.

31. 40% of _40_ is 100.

32. 10% of _17_ is 1.7.

33. 10% of 1.7 is _17_.

34. 25% of _2.5_ is 10.

Solve these word problems.

35. Barclay Stokes gets an annual salary of $25,000. He and his family spend $3,500 per year on food. What percent of his salary is spent for food?

Answer _____

36. The Economy Appliance Store has a television set that regularly sells for $249. If the store offers to sell the set at a 22% discount, what would be the sale price?

Answer _____

27

Problem-Solving Strategy: Multi-Step Problems

Solving a problem requires several steps such as reading the problem, deciding what to do, and finding the answer. Some steps of solving the problem may need to be broken down into *substeps*, or smaller steps. Identifying the substeps is important in planning how to solve the main problem.

STEPS
1. **Read the problem.**

   In one week Wendy worked 42.5 hours. She earned overtime for each hour over 40. Her overtime rate was 1.5 times her regular rate of $8.70 per hour. What was her gross pay for the week?

2. **Identify substeps.**
   1. How many hours of overtime pay did Wendy earn?
   2. How much was she paid for each hour of overtime?
   3. How much was the total overtime pay?
   4. How much was the total pay for the 40 regular hours?

3. **Solve the problem.**

   First find the answer for each substep.
   1. 42.5 − 40 = 2.5 hours overtime
   2. 1.5 × $8.70 = $13.05 per hour of overtime
   3. 2.5 × $13.05 = $32.625 or $32.63 total overtime pay
   4. 40 × $8.70 = $348.00 total regular pay

   Then solve the main problem.

Regular pay		Overtime pay		Gross pay
$348.00	+	$32.63	=	$380.63

   Wendy's gross pay for the week was $380.63.

Solve. Use substeps.

1. When Randy started his job, he earned $5.40 an hour. Now he makes $234 for a 40-hour week. How much more does Randy earn per hour than he did when he started?

   Answer $0.45

2. Evelyn earns a weekly salary of $560. Wanda earns a monthly salary of $2,370. Who earns more per year? How much more?

   Answer Evelyn earns more $680

3. Xavier earns a yearly salary of $16,500. If he gets a raise of $110 per month, what is the percent of increase?

   Answer 8%

4. In one week, Quenton earned $72.90 for overtime hours. He earns time-and-a-half for overtime. If his regular rate is $8.10 per hour, how many hours of overtime did Quenton work?

   Answer 6 hrs

28

Applying Your Skills

handwritten: 376 405   *handwritten: I = prt*

Solve.

1. If you received a score of 80% on a test of 20 problems, how many problems did you have correct?

Answer ___16___

2. Edith bought a $62 coat at 20% discount. How much did she pay for the coat?

Answer $12.40

3. Alfredo Gonzales borrowed $450 at 6% annual interest for 8 months. How much interest did he owe?

Answer $18.00

4. Sharon repaid a $500 loan 16 months after it was made. At 6% annual interest, how much interest did she owe?

Answer $40.00

5. Sol bought 4 shirts on sale at 25% off. They had been marked $25 each. How much did he save on the 4 shirts? (First, find how much was saved on one shirt. Then, multiply by 4.)

Answer $11.25   *he saved on each shirt*

6. By paying cash, Mrs. Brewer received a discount of 5% on the purchase of $750 in furniture. How much did she save?

Answer $37.50

7. Mirna bought a suit on sale. The suit had been marked to sell for $139.00. How much did she save by buying at 40% off?

Answer $55.60

8. A rancher had 1,500 acres of grazing land for her cattle. This represented 30% of the entire ranch. How many acres are there in the ranch?

Answer 1,950 acres

9. Jackson bought a table at a discount of 15%. The original price of the table was $75. How much did he save on the table?

Answer $11.25

10. At 5% simple interest, Mamie put $300 in a savings account on the first day of the year. How much would she have in her savings account on the last day of the year?

Answer $114,975.00   *315*

11. Bonnie got a raise of 32¢ per hour. Now she earns $314.40 for a 40-hour week. What was her hourly rate before her raise?

Answer $7.54

12. One week, Christine earned 8% commission on sales of $4,700. The next week, she earned 9% commission on sales of $5,500. How much more commission did she earn the second week than the first week?

Answer 119 CS.   *300 .05 / 15.00*

Change each percent to a decimal and then to a fraction. Simplify.

1. 0.4% = _.04_   35.5% = _35.5_   11% = _.11_   250% = _.250_

Change the following to percents.

2. $\frac{7}{8}$ = _100_   3.45 = _3.45%_   0.003 = _0003_ $\frac{4}{25}$ = _1.500_

Find each percent or number.

3. 54% of 75 is _40.50_.   0.9% of 200 is _180_.   7% of 35 is _2.45_.

4. _16_% of 25 is 4?   _5_% of 18 is 36?   25% of _68_ is 17?

Solve.

---

5. How much will a $160 stereo cost if it is on sale for 30% off?

Answer _112_

---

6. Sales tax is 6.5%. What will the final cost be on a shirt that costs $19.99? Round to the nearest cent.

Answer _$129.94_

---

7. Mrs. Lynch bought a suit which had sold for $80. She bought it at 25% discount. How much did she save?

Answer _$60.00 She saved $20.00_

---

8. A real estate agent rented a store to a grocer for $1,000 per month, charging a commission of 5%. What was the net rental received by the owner?

Answer _1,050_

---

9. Sally Rosenthall borrowed $450 at the bank for 3 months. The interest rate was 4%. How much interest was due on the loan?

I = p r t

450 4% 3 $\frac{3}{12}$

Answer _468_   intrest was $18

---

10. Myrl was making $6 per hour and was given a 7% raise. How much does she make per hour with the increase?

Answer _$6.42_

Unit 2   Measurement and Formulas

Time

The next few pages are devoted to using tables of measures and to converting from one unit to another.

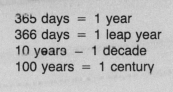

60 seconds (sec) = 1 minute (min)	365 days = 1 year
24 hours (hr) = 1 day	366 days = 1 leap year
7 days = 1 week (wk)	10 years = 1 decade
52 weeks = 1 year	100 years = 1 century
12 months (mo) = 1 year (yr)	

To change from a larger unit of time to a smaller unit of time, multiply the units. To change from a smaller unit of time to a larger unit of time, divide the units.

Find: 7 days = _____ hours

To change a larger unit to a smaller unit, multiply.

24 hours = 1 day

$$\begin{array}{r} 2\,4 \text{ hours} \\ \times\ 7 \text{ days} \\ \hline 1\,6\,8 \text{ hours} \end{array}$$ 7 days = 168 hours

Find: 441 days = _____ weeks

To change a smaller unit to a larger unit, divide.

7 days = 1 week

$\frac{441 \text{ days}}{7 \text{ days}} = 63$

441 days = 63 weeks

Change each larger unit to a smaller unit. You may need to change units twice.

1. 5 yr = __260__ wk   30 wk = __210__ days   17 yr = __884__ wk   13 min = __780__ sec

2. 5 yr = __1825__ days   3 hr = __10,800__ sec   72 days = __1728__ hr   2 yr = __17520__ hr

Change each smaller unit to a larger unit. You may need to change units twice.

3. 300 yr = __30__ decades   48 hr = __2__ days   18 mo = __1½__ yr   1,095 days = __3__ yr

4. 7,200 sec = __2__ hr   84 days = __12__ wk   84 hr = __3.5__ days   286 wk = __5.5__ yr

Solve.

5. Maria works out for 150 minutes every Saturday. How many hours is this?

Answer _____

6. Wong ran for 45 minutes in a marathon without stopping. How many seconds did Wong run without stopping?

Answer _____

Customary Length

The common customary units that are used to measure length are inch, foot, yard, and mile. The chart shows a list of common measures.

1 foot (ft) = 12 inches (in.)	1 yard (yd) = 3 feet = 36 inches	1 mile (mi) = 1,760 yards = 5,280 feet

Find: $3\frac{1}{2}$ ft = _____ in.

> To change feet to a smaller unit, multiply.
> Use fractions.
>
> 1 ft = 12 in.
>
> $3\frac{1}{2} \times 12 = \frac{7}{2} \times \frac{\cancel{12}^{6}}{1} = \frac{42}{1} = 42$
>
> $3\frac{1}{2}$ ft = 42 in.

Find: 10 ft = _____ yd

> To change feet to a larger unit, divide.
>
> 3 ft = 1 yd
>
> $\frac{10\,ft}{3\,ft} = 3\frac{1}{3}$
>
> 10 ft = $3\frac{1}{3}$ yd

Change each measurement to the smaller unit. For some problems, you may need to change units twice.

1. $5\frac{1}{6}$ yd = __$15\frac{1}{2}$__ ft $1\frac{1}{2}$ mi = __7920__ ft $7\frac{1}{3}$ ft = __88__ in.   2 mi = __126720__ in.

2. $1\frac{1}{4}$ yd = __~~180~~ 45__ (in.)   21 yd = __63__ ft   4 mi = __7040__ yd   4 yd = __144__ in.

Change each measurement to the larger unit. For some problems, you may need to change units twice.

3. 51 in. = __$4\frac{1}{4}$__ ft   38 ft = __$12\frac{2}{3}$__ yd   6 in. = __$\frac{1}{2}$__ ft   440 yd = __$\frac{1}{4}$__ mi

4. 4,400 yd = __2.5__ mi   222 in. = __6__ yd   15,840 ft = __3__ mi   24 in. = __1.5__ yd

Solve.

5. Beth used 24 inches of material to make a pillow cover. How many yards is this?

    Answer _____

6. The distance from José's house to the library is $\frac{1}{2}$ mile. How many yards is this?

    Answer _____

Customary Weight

The customary units that are used to measure weight are ounce, pound, and ton. The chart shows a list of common measures.

<table>
<tr><td>1 pound (lb) = 16 ounces (oz)</td></tr>
<tr><td>1 ton (T) = 2,000 pounds</td></tr>
</table>

Find: $5\frac{1}{2}$ lb = _____ oz

Find: 6,500 lb = _____ T

To change pounds to a smaller unit, multiply. Use fractions.

$$1 \text{ lb} = 16 \text{ oz}$$

$$5\frac{1}{2} \times 16 = \frac{11}{2} \times \frac{\overset{8}{16}}{1} = \frac{88}{1} = 88$$

$$5\frac{1}{2} \text{ lb} = 88 \text{ oz}$$

To change pounds to a larger unit, divide.

$$2,000 \text{ lb} = 1 \text{ T}$$

$$\frac{6,500}{2,000} = \frac{65}{20} = 3\frac{5}{20} = 3\frac{1}{4}$$

$$6,500 \text{ lb} = 3\frac{1}{4} \text{ T}$$

Change each measurement to the smaller unit. For some problems, you may need to change units twice.

1. $3\frac{1}{4}$ lb = __52__ oz $2\frac{1}{2}$ T = __5,000__ lb   6 lb = __96__ oz   1 T = __32,000__ oz

2. 4 T = __8,000__ lb $1\frac{3}{8}$ lb = __✗ 88__ oz $4\frac{3}{16}$ lb = __67__ oz $\frac{1}{2}$ T = __1,000__ lb

Change each measurement to the larger unit. For some problems, you may need to change units twice.

3. 53 oz = __$3\frac{5}{16}$__ lb   7,000 lb = __3.5__ T   80 oz = __5__ lb   800 lb = __✗ 2.5__ T

4. 72 oz = __✗ 5__ lb   36 oz = __3.25__ lb   2,400 lb = __1.2__ T   32,000 oz = __16 ✗__ T

Solve.

5. Andrea bought a beef roast which weighed $14\frac{1}{4}$ pounds. How many ounces is this?

Answer _____

6. Conrad bought $5\frac{1}{2}$ tons of coal. How many pounds is this?

$$16\overline{\smash{)}30} \\ \underline{-16}$$

01

16
× 3

Answer _____

Customary Capacity

The customary units that are used to measure capacity are fluid ounce, cup, pint, quart, and gallon. The charts show a list of common measures.

4/29

LIQUID

8 fluid ounces (fl oz)	= 1 cup (c)
1 pint (pt)	= 2 cups
1 quart (qt)	= 2 pints
	= 4 cups
1 gallon (gal)	= 4 quarts
	= 8 pints
	= 16 cups

DRY

2 pints	= 1 quart
8 quarts	= 1 peck (pk)
4 pecks	= 1 bushel (bu)

Find: $3\frac{3}{4}$ qt = _____ c

To change quarts to a smaller unit, multiply. Use fractions.

$$1 \text{ qt} = 4 \text{ c}$$

$$3\frac{3}{4} \times 4 = \frac{15}{4} \times \frac{4}{1} = \frac{15}{1} = 15$$

$$3\frac{3}{4} \text{ qt} = 15 \text{ c}$$

Find: 21 qt = _____ gal

To change quarts to a larger unit, divide.

$$4 \text{ qt} = 1 \text{ gal}$$

$$\frac{21}{4} = 5\frac{1}{4}$$

$$21 \text{ qt} = 5\frac{1}{4} \text{ gal}$$

Change each measurement to the smaller unit. For some problems, you may need to change units twice.

1. $8\frac{1}{2}$ pt = _**17**_ c $2\frac{1}{4}$ gal = _**9**_ qt   13 qt = _**26**_ pt   2 bu = _**64**_ qt

2. $4\frac{1}{8}$ gal = _**33**_ pt   9 qt = _**36**_ c $1\frac{1}{2}$ gal = _**24**_ c   30 pk = _**240**_ qt

Change each measurement to the larger unit. For some problems, you may need to change units twice.

3. 7 c = _$1\frac{3}{4}$_ qt   21 pt = _**3**_ gal   *2⅝*   11 c = _$5\frac{1}{2}$_ pt   16 pk = _**4**_ bu

4. 24 qt = _**6**_ gal   14 pt = _**7**_ qt   54 c = _$3\frac{3}{8}$_ gal   20 qt = _$2\frac{1}{2}$_ pk

Solve.

5. Antonio made 8 gallons of spaghetti sauce. How many pint jars can he fill?

   Answer _____

6. The juice from six large oranges measures about three cups. How many pints is this?

   Answer _____

34

You can add, subtract, multiply, and divide measures that are given in two units. For example, 1 yd 9 in. uses 2 units to measure one length.

Find: 1 lb 14 oz + 6 lb 10 oz

Add.	Change 24 ounces into pounds and ounces.
1 lb 14 oz + 6 lb 10 oz 7 lb 24 oz	$\frac{24\ oz}{16\ oz} = 1\ lb\ 8\ oz$ Add 1 lb 8 oz to 7 lb.
	7 lb + 1 lb 8 oz = 8 lb 8 oz

Find: 5 × 4 gal 3 qt

Multiply.	Change 15 quarts into gallons and quarts.
4 gal 3 qt × 5 20 gal 15 qt	$\frac{15\ qt}{4\ qt} = 3\ gal\ 3\ qt$ Add 3 gal 3 qt to 20 gal.
	20 gal + 3 gal 3 qt = 23 gal 3 qt

Find: 12 ft 4 in. − 6 ft 9 in.

Subtract. You cannot subtract 9 in. from 4 in.	Change 12 ft into feet and inches. 12 ft = 11 ft 12 in. Add 11 ft 12 in. to 4 in. 11 ft 12 in. + 4 in. = 11 ft 16 in.
12 ft 4 in. → − 6 ft 9 in. →	11 ft 16 in. − 6 ft 9 in. 5 ft 7 in.

Find: 8 T 200 lb ÷ 6

Divide 8 T by 6. Subtract.	Change 2 T 200 lb to pounds.
1 7 700 lb 6)8 7 200 lb 6 2 7 200 lb 4,200 4,200 0	2 T × 2,000 = 4,000 lb 4,000 lb + 200 lb = 4,200 lb Divide 4,200 lb by 6. Subtract.

Change each unit.

1. 30 ounces = __1__ lb __14__ oz   18 inches = __1__ ft __6__ in.   12 gallons = 11 gal __4__ qt

Find each answer.

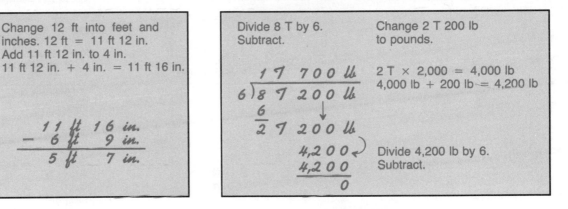

2.
```
  5 gal  2 qt          9 ft  6 in.          7 yd  1 ft
+ 2 gal  3 qt        + 4 ft  8 in.        + 3 yd  2 ft
```
7 gal 5 qt = 8 gal 1 qt   14 ft 2 in   11 yd

3.
```
  8 yd  7 ft          5 gal  2 qt         24 lb   8 oz
- 6 yd  2 ft        - 3 gal  3 qt       - 10 lb  12 oz
```
7 yd 2 ft   1 gal 3 qt   13 lb 12 oz

4.
```
 18 lb  6 oz          5 ft  8 in.         6 T 500 lb
×       4           ×        7           ×         8
```
73 lb 8 oz   39 ft 8 in.   50 T

5. 3)5 gal 1 qt   8)10 lb 8 oz   7)9 yd 1 ft

35

Practice in Computing Measures

Add the following quantities.

1.
```
   1 2 ft 1 0 in.          5 lb   6 oz          7 hr 1 5 min
 +   5 ft   4 in.        + 4 lb 1 0 oz        + 4 hr 4 5 min
```

2.
```
   2 yd 2 ft              3 T     5 0 0 lb      2 mi 1,2 0 0 yd
 + 3 yd 2 ft            + 4 T 1,5 0 0 lb      + 7 mi   7 0 0 yd
```

Subtract.

3.
```
   5 yd 2 ft              3 qt 1 pt            1 8 lb 2 oz
 - 2 yd 1 ft            - 2 qt              - 1 4 lb 6 oz
```

4.
```
   5 gal 3 qt             5 T 1,0 0 0 lb      1 5 ft 9 in.
 - 2 gal 2 qt           - 2 T 1,6 0 0 lb    -   5 ft 4 in.
```

Multiply each of these measures.

5.
```
   4 lb 2 oz             5 T 8 0 0 lb         3 ft 9 in.
 ×       6             ×           5        ×         5
```

6.
```
   3 gal 3 qt            2 yd 2 ft            6 qt 1 pt
 ×         4            ×       3            ×       4
```

Divide.

7. 3)8 min 6 sec   3)4 T 1 0 0 lb   5)8 yd 1 ft

8. 7)1 0 qt 1 pt   4)5 days 8 hr   6)8 lb 4 oz

Solve.

9. Pablo is 20 yr 8 mo old. His sister is 2 yr 10 mo younger. How old is Pablo's sister?

Answer _____

10. A bolt of cloth containing 25 yd 6 in. was cut into 6 pieces. How long was each piece?

Answer _____

Metric Length

The meter (m) is the basic metric unit of length. A meter can be measured with a meter stick. The length of your arm is about 0.7 m.

A centimeter (cm) is one hundredth of a meter. (Centi means 0.01.) The centimeter is used to measure small lengths. The thickness of a nickel is about 0.2 cm.

A millimeter (mm) is one thousandth of a meter. (Milli means 0.001.) The millimeter is used to measure very small lengths. The thickness of a nickel is about 2 mm.

A kilometer (km) is one thousand meters. (Kilo means 1,000.) The kilometer is used to measure long distances. The distance between two cities is 98 km.

1 km	=	1,000 m
1 m	=	100 cm
1 cm	=	10 mm

1 m	=	0.001 km
1 cm	=	0.01 m
1 mm	=	0.1 cm

Find: 8.2 m = _____ cm

> To change meters to a smaller unit, multiply.
>
> $$1 \text{ m} = 100 \text{ cm}$$
> $$8.2 \times 100 = 820$$
> $$8.2 \text{ m} = 820 \text{ cm}$$

Find: 63 m = _____ km

> To change meters to a larger unit, divide.
>
> $$1,000 \text{ m} = 1 \text{ km}$$
> $$\frac{63}{1000} = 0.063$$
> $$63 \text{ m} = 0.063 \text{ km}$$

Change each measurement to the smaller unit.

1. 14.5 km = _14,500_ m   7.25 m = _725._ cm   18 cm = _180_ mm   6.5 km = _6500.0_ m

2. 3.4 m = _340._ cm   21 m = _210._ mm _21,000_   0.9 km = _0900._ m   29 cm = _290._ mm

Change each measurement to the larger unit.

3. 48 mm = _4.8_ cm   79.6 cm = _.796_ m   61 m = _.061_ km   657 mm = _.657_ m

4. 8,542 m = _8.542_ km   3,128 mm = _3.128_ m   930 cm = _9.30_ m   75 mm = _7.5_ cm

Solve.

5. Korinne had 15 meters of fabric. How many centimeters of fabric does she have?

Answer _____

6. The Bakers traveled 32,500 meters to see their cousins. How many kilometers did they travel?

Answer _____

37

The word *mass* is not often used outside the field of science. The common term for mass is weight.

The gram (g) is the basic metric unit of mass. The gram is used to measure the weight of very light objects. A dime weighs about 2 grams.

The kilogram (kg) is one thousand grams. It is used to measure the weight of heavier objects. Use kg for the weight of a computer. Remember, kilo means 1,000.

$$1 \text{ kg} = 1,000 \text{ g}$$

$$1 \text{ g} = 0.001 \text{ kg}$$

Find: 8.4 kg = _____ g

To change kilograms to a smaller unit, multiply.

$$1 \text{ kg} = 1,000 \text{ g}$$
$$8.4 \times 1,000 = 8,400$$
$$8.4 \text{ kg} = 8,400 \text{ g}$$

Find: 24.7 g = _____ kg

To change grams to a larger unit, divide.

$$1,000 \text{ g} = 1 \text{ kg}$$
$$\frac{24.7}{1000} = 0.0247$$
$$24.7 \text{ g} = 0.0247 \text{ kg}$$

Change each measurement to the smaller unit.

1. 32 kg = _32,000_ g   0.007 kg = _____ g   1.8 kg = _____ g   526 kg = _____ g

2. 0.49 kg = _____ g   825 kg = _____ g   6.783 kg = _____ g   0.08 kg = _____ g

Change each measurement to the larger unit.

3. 12.8 g = _0.0128_ kg   9 g = _____ kg   137 g = _____ kg   23.25 g = _____ kg

4. 5,268 g = _____ kg   25 g = _____ kg   4.9 g = _____ kg   789 g = _____ kg

Solve.

5. A bag of potatoes holds 25 kilograms. How many grams of potatoes are in the bag?

   Answer _____

6. A carton of juice weighs 246 grams. How many kilograms does the carton weigh?

   Answer _____

Metric Capacity

The liter (L) is the basic metric unit of capacity. A liter of liquid will fill a box 10 centimeters on each side. A large jug of apple cider holds about 4L.

A milliliter (mL) is one thousandth of a liter. It is used to measure very small amounts of liquid. A milliliter of liquid will fill a box 1 centimeter on each side. A small carton of milk holds about 250 mL.

Remember, milli means 0.001.

1 cm
1 cm 1 cm

1 L = 1,000 mL		1 mL = 0.001 L

Find: 2.5 L = _____ mL

To change liters to a smaller unit, multiply.

1 L = 1,000 mL

2.5 × 1,000 = 2,500

2.5 ℓ = 2,500 mℓ

Find: 5,672 mL = _____ L

To change milliliters to a larger unit, divide.

1,000 mL = 1 L

$\frac{5672}{1000} = 5.672$

5,672 mℓ = 5.672 ℓ

Change each measurement to the smaller unit.

1. 27 L = _27,000_ mL   5.3 L = _____ mL   7.45 L = _____ mL   0.5 L = _____ mL

2. 0.825 L = _____ mL   2 L = _____ mL   39.6 L = _____ mL   2.6 L = _____ mL

Change each measurement to the larger unit.

3. 3,096 mL = _3.096_ L   6,000 mL = _____ L   412.5 mL = _____ L   528 mL = _____ L

4. 58 mL = _____ L   798 mL = _____ L   19.2 mL = _____ L   6.34 mL = _____ L

Solve.

5. How many milliliters of medicine are in a 4.5 liter container?

Answer _____

6. A measuring cup holds 250 milliliters. How many liters does the measuring cup hold?

Answer _____

Checking Up

Change each of these measurements as indicated.

1. 3 in. = **.4** ft   49 in. = **1 13/36** yd

2. 80 ft = **26 2/3** yd   1⅓ pt = ✗ **2 2/3** c

3. 64 m = **0.064.** ✗ km   18 sec = **.3** min

4. 5 pk = **1 1/4** bu   86.09 m = **8609** cm

5. 57 lb = **912** oz   1 kg = **1,000** g

6. 1 mL = ✗ **0.001** L   1 mi = **63360** in.

7. 65.3 kg = **65,300** g   ½ fl oz = **.6** c

Add.

8.
~~15~~ **16** min 1 5 sec
+ 2 5 min 4 5 sec
41 min

3 hr 2 5 min
+ 2 hr 2 5 min
5 hr 50 min

~~7~~ **8** days 1 2 hr
+ 5 days 1 8 hr
13 days 6 hr

Subtract.

9.
6
~~5~~ ft   7 in.
− 2 ft 1 0 in.
8 ft 5 in

3 4
4 yd ~~1~~ ft
− 2 yd 2 ft
1 yd 2 ft

5 T   6 0 0 lb
− 3 T 1,2 0 0 lb
8 T 1,800 lb

Multiply.

10.
6 hr 1 ~~5~~ min
×   4
25 hr 0

3
6 bu 3 pk
×   4
27 bu

2
2 T 8 0 0 lb
×   6
14 T 04

Divide.

11. 4)1 3 hr 2 0 min   7)1 0 pk 4 qt   6)7 days 6 hr

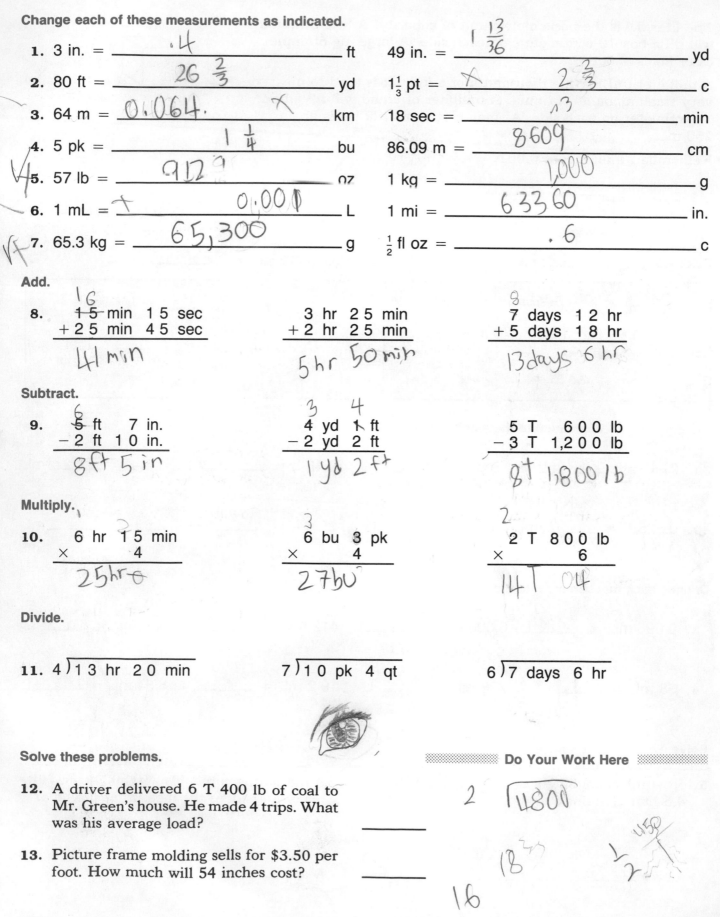

Solve these problems.

12. A driver delivered 6 T 400 lb of coal to Mr. Green's house. He made 4 trips. What was his average load? _____

2)4800

13. Picture frame molding sells for $3.50 per foot. How much will 54 inches cost? _____

18 ³⁵

450

1/2

16

40

Using Measures

$\frac{4}{3} \times \frac{2}{1}$ $8\overline{)80}$ $60\overline{)18\cdot0}$

Solve these mixed-measure problems.

1. Mrs. Ortiz had a roll of shelf paper 9 meters long. She cut off a piece for the shelf 2.1 meters long. How much did she have left in the roll?

   Answer _____ 2 _____

2. A trucker had a load of coal containing 2 tons. She unloaded 1 T 1,200 lb. How much did she have left on the truck?

   Answer _____ ,800 lb _____

3. Thomas is 34 yr 7 mo old. His sister is 2 yr 10 mo younger. How old is the sister?

   Answer _____ 31 yrs 9 mo _____

4. I am 34 yr 6 mo old. My daughter is 7 yr 9 mo old. I am how much older than my daughter?

   Answer _____ 26 yrs 9 mo _____

5. Miss Jones bought a chicken weighing 4 lb 12 oz at $0.96 per lb. What was the total cost?

   Answer _____ $4.56 _____

6. One year we bought 8 T 1,200 lb of coal. At $40 per ton, how much did it cost?

   Answer _____ $344 _____

7. Mr. Lee bought a piece of material which contained 5 yards and 18 inches. If he paid $4.50 per yard, how much did it cost?

   Answer _____ $24.75 _____ $22.50

8. Mr. Green had 6 yds 9 in. of flower bed edging. He cut it into 5 pieces of the same length. How long was each piece?

   Answer _____ 1 yd 9 in _____

9. A load of coal weighing 3 T 600 lbs is to be unloaded, one-half at one place and the remainder at another. How much is to be left in each place?

   Answer _____ 1 T 800 lb _____

10. Three brothers weigh 176 lb 10 oz, 184 lb 8 oz, and 196 lb 2 oz. What is their average weight?

    1 Lb 4 oz

    Answer _____ 185 lb 8 oz _____

11. Murray's best record for the long jump was 12 ft 8 in. Stanton's best jump was 11 ft 11 in. How much better was Murray's jump?

    Answer _____ 9 in _____

12. It took 7 hr 30 min for us to drive the last 375 miles of our trip. How many miles per hour did we average?

    Answer _____

41

Drawing to Scale

Scale drawings are used to represent the actual dimensions of objects and distances. A map is an example of a scale drawing representing distances. Maps have a scale that shows distance between points.

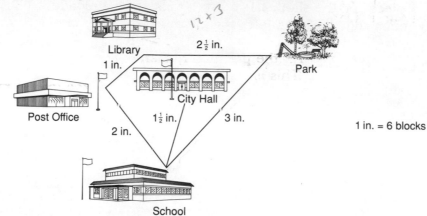

1 in. = 6 blocks

Solve the following using the map above.

1. If 1 inch represents 6 blocks, how many blocks apart are the library and the park?

   Answer _____ 15 blocks _____

2. If 1 inch represents 6 blocks, how many blocks is it from city hall to the school?

   Answer ___ 9 blocks ___

3. If 1 inch represents 6 blocks, how many blocks is the school from the park?

   Answer ___ 18 blocks ___

4. If 1 inch represents 6 blocks, how many blocks is it from the school to the library if you go past the post office?

   Answer ___ 18 blocks ___

5. If 1 inch represents 6 blocks, how many blocks is it from the post office to the park if you go past the school?

   Answer ___ 30 blocks ___

6. A race starts at the school and goes past the park, the library, the post office, and ends back at the school. How many blocks will the race cover?

   Answer ___ 51 blocks ___

Solve the word problems below.

7. If 1 inch represents 25 miles on a road map, how far apart are 2 cities that are 7 inches apart on the map?

   Answer ___ 175 ___

8. In a scale drawing, if 1 inch represents 12 feet, how many feet are represented by $2\frac{1}{2}$ inches?

   Answer ___ 30 ___

9. A geography map has this notation: Scale: 1 in. = 150 mi. If New York is 12 inches from Denver on the map, how far apart are these cities?

   Answer ___ 1,800 ___

10. If 1 inch represents 50 miles, how many inches will represent 500 miles?

    Answer _____

Formula for Perimeter of a Rectangle

The figure at the right is a **rectangle.** The opposite sides of a rectangle are equal. The opposite sides are also **parallel,** which means that they are the same distance apart at all points.

To find the **perimeter** of a rectangle, add all four sides. If we let l represent the length, w the width and P the perimeter, then $P = l + w + l + w$ or $P = 2l + 2w$. This is a formula.

EXAMPLE: Find the perimeter of a rectangle that measures 40 m by 60 m.

$w = 40$ m

$l = 60$ m

Write the formula. $\quad P = 2l + 2w$

Substitute the data. $\quad P = (2 \times 60) + (2 \times 40)$

Solve the problem. $\quad P = 120 + 80$

$\qquad\qquad\qquad\quad P = 200$

The perimeter of the rectangle is 200 meters.

Use the formula for perimeter of a rectangle. Solve.

---

1. Ned wants to fence his garden. It is 15 meters wide and 20 meters long. How much fencing does he need?

Answer ___70___

2. Harvey Park is 0.25 kilometer wide and 0.175 kilometer long. What is the perimeter of the park?

Answer ___0.850___

---

3. A pillow measures 54 centimeters by 62 centimeters. How much braid is needed to go around the pillow?

Answer ___232___

4. A desk measures 159 centimeters by 120 centimeters. What is the perimeter of the desk?

Answer ___558___

---

5. A farm measures 1.4 kilometers by 1.2 kilometers. What is the perimeter of the farm?

Answer ___5.2___

6. A rectangle is twice as long as it is wide. The rectangle is 20.5 millimeters wide. What is the perimeter of the rectangle?

Answer ___123___

Formula for Area of a Rectangle

The surface enclosed by the lines of a plane figure is called the **area.** The area of any figure is the number of *square units* it contains.

The formula, $A = lw$, means the area of a rectangle equals the length times the width.

EXAMPLE: Find the area of a rectangle that measures 12 m by 8 m.

Write the formula. $A = l \times w$

Substitute the data. $A = 12 \times 8$

Solve the problem. $A = 96$

The area of the rectangle is 96 square meters.

Remember, write your answer in *square* units.

$w = 8\,m$

$l = 12\,m$

Use the formula for area of a rectangle. Solve.

---

1. How many square meters are in the floor area of a room that measures 8 meters by 6.9 meters?

   Answer ___55.2___

2. What is the area of a window that measures 16 inches by 24 inches?

   Answer ___384___

---

3. A pillow measures 62 centimeters by 40 centimeters. How much fabric will be needed to cover one side of the pillow?

   Answer ___2,480___

4. How much sod is needed to cover a yard that measures 20 meters by 15 meters?

   Answer ___300___

---

5. What is the area of a concrete walk that measures 100 meters by 1.5 meters?

   Answer ___150___

6. How much linoleum is needed to cover a floor that measures 54 feet by 42 feet?

   Answer ___2,268___

---

7. How many square inches of glass are needed to cover a picture that measures 8 inches by 10 inches?

   Answer ___80___

8. How much carpeting is needed to cover a floor that measures 3 meters by 4.4 meters?

   Answer ___13.2___

Formulas for Area and Perimeter of a Square

A **square** is a rectangle with all sides equal. This means the length and width of a square are the same. To find the area and perimeter of a square, you can use formulas.

The formula $A = s \times s$ means the area of a square equals the side times the side.

EXAMPLE: Find the area of a square with one side equal to 8 inches.

Write the formula. $A = s \times s$

Substitute the data. $A = 8 \times 8$

Solve the problem. $A = 64$

The area of this square is 64 square inches.

s = 8 inches

The formula $P = 4s$ means the perimeter of a square equals 4 times the side, because all the sides of a square are equal.

EXAMPLE: Find the perimeter of the same square.

Write the formula. $P = 4s$

Substitute the data. $P = 4 \times 8$

Solve the problem. $P = 32$

The perimeter of this square is 32 inches.

Remember, only area is written in square units.

Use a formula to find the area and the perimeter of each square.

1.

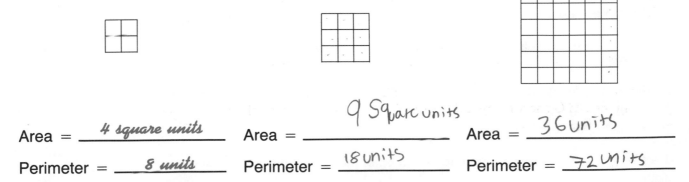

Area = __4 square units__   Area = __9 Square units__   Area = __36 units__

Perimeter = __8 units__   Perimeter = __18 units__   Perimeter = __72 units__

Use the formula for area or perimeter of a square. Solve.

2. Yolanda wants to put a new carpet in her bedroom. The bedroom is in the shape of a square that measures 12 feet on each side. How many square feet of carpet will Yolanda need?

Answer __24__

3. Andre wants to put a fence around his square garden. Each side of the garden is 15 feet long. How many feet of fencing will Andre need?

Answer __30__

You know that area measures the surface of a figure. **Volume** is the space enclosed (capacity) within a three-dimensional figure, such as a bin, a box, or a room. Area is measured in square units. **Volume is measured in *cubic units*.** The exercises below will help you understand cubic measurement.

At the right is a drawing of the ice-cube tray. The cubes are an inch long, an inch wide, and an inch thick (or deep).

Answer these questions.

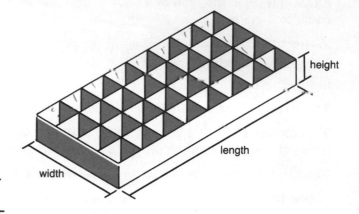

1. How many cubes deep is the tray? __1__

2. How many inches deep is the tray? __1__

3. How many cubes wide is the tray? __4__

4. How many inches wide is it? __4__

5. How many cubes are in the length? __8__

6. How many inches are in the length? __8__

7. How many cubes are in the entire tray? __32__

8. Multiply the length, width, and depth (or height) of the tray.

   How many cubic inches are in it? __32__ Does this agree

   with the number of ice cubes found in problem 7? __yes__

The formula $V = lwh$ is used to find the volume of a rectangular solid (box, bin, room). The formula means the volume equals the length times the width times the height.

EXAMPLE: Find the volume of a box which measures 10 m by 15 m by 4 m.

$V = l \times w \times h$
$V = 15 \times 10 \times 4$
$V = 600 \ cubic \ meters$

(box diagram: h = 4 m, w = 10 m, l = 15 m)

Use the formula, assume all dimensions to be in the same unit, and complete the charts.

9.

l	*w*	*h*	*V*
10	20	5	1,000
25	24	10	6,006
34	28	16	15,232
1.5	4	2.5	15
$12\frac{1}{2}$	10	20	2,500

10.

l	*w*	*h*	*V*
12	15	8	1,440
16	20	6	1,920
22	24	12	6,336
5.5	8	6.4	281.6
$8\frac{1}{2}$	12	10	1,020

Lines and Angles

Lines and **angles** makes up **geometric shapes.** Study these examples to prepare for geometric problems.

Lines

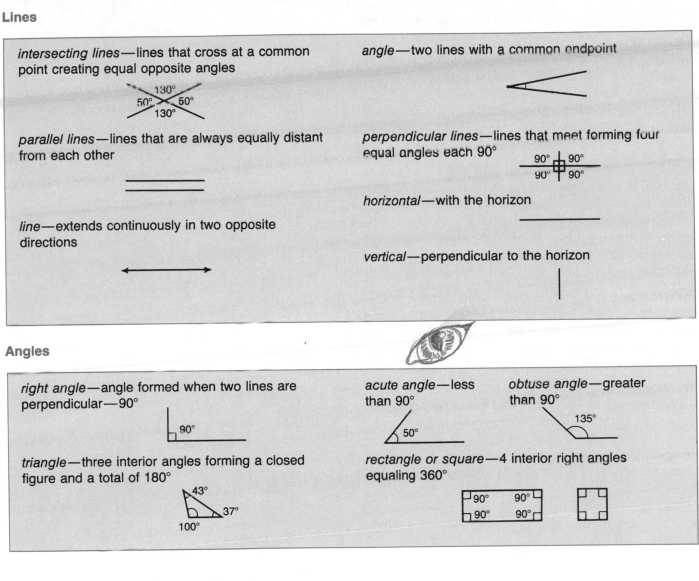

intersecting lines—lines that cross at a common point creating equal opposite angles

angle—two lines with a common endpoint

parallel lines—lines that are always equally distant from each other

perpendicular lines—lines that meet forming four equal angles each 90°

line—extends continuously in two opposite directions

horizontal—with the horizon

vertical—perpendicular to the horizon

Angles

right angle—angle formed when two lines are perpendicular—90°

acute angle—less than 90°

obtuse angle—greater than 90°

triangle—three interior angles forming a closed figure and a total of 180°

rectangle or square—4 interior right angles equaling 360°

Write the name for each type of line.

1. _parallel_

2. Perpendicular

3. Line

120°
60° 60°
120° intersecting

vertical

angle

Write the name for each type of angle.

4. 45° _acute_

square.

5. 60° / 60° 60°

105° obtuse

47

The **triangle** (△) is a closed, three-sided geometric figure. Each of the three points where an angle is formed is called a **vertex.** Each vertex of the triangle is formed when two sides meet to form an angle.

In triangle RST (△RST), the vertices (plural of vertex) are points R, S, and T. The sides of the triangle are RS and ST. RT is the base of the triangle. The angles of △RST are ∠R, ∠S, and ∠T. Sometimes we use three letters to identify angles of a triangle. ∠RST, ∠STR, and ∠TRS are the angles of the triangle. The middle letter is the angle being identified when we use three letters.

Triangles are classified by referring to their sides or their angles. When the three sides are equal, the triangle is called an **equilateral** triangle. When two sides of the triangle are equal, the triangle is called an **isosceles** triangle. If all three sides have different lengths, the triangle is called a **scalene** triangle.

An **acute** triangle has three acute angles. The **obtuse** triangle has one obtuse angle. The **right** triangle has one right angle. An **equiangular** triangle has three equal angles and is also an **equilateral** triangle with all sides equal.

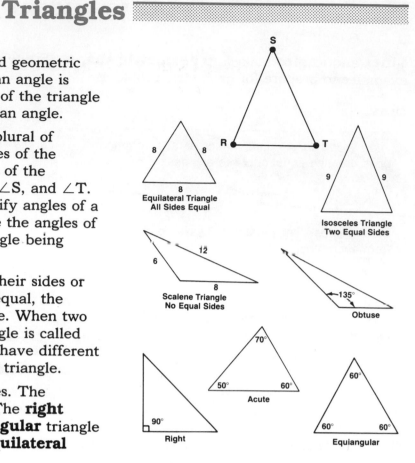

Label the following triangles as equilateral, isosceles, or scalene.

1.

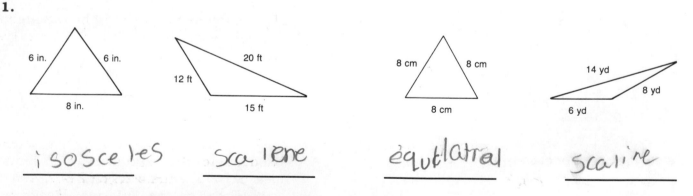

isosceles Scalene equtlatral scaline

Label the following triangles as acute, obtuse, right, or equiangular.

2.

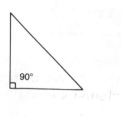 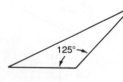

right obtusc equinagular acute

Formula for Perimeter of a Triangle

The perimeter of any triangle is the sum of its three sides.

EXAMPLE: Find the perimeter of an equilateral triangle, with one side equal to 24 feet.

$P = s_1 + s_2 + s_3$
$P = 24 + 24 + 24$
$P = 72$

The perimeter is 72 feet.

Use what you know about the kinds of triangles to solve these problems.

1. In the **isosceles** triangle EFG, EG = GF = $6\frac{1}{2}$ ft and EF = $3\frac{1}{2}$ ft. Find the perimeter of triangle EFG.

$P = 6\frac{1}{2} + 6\frac{1}{2} + 3\frac{1}{2} = 16\frac{1}{2}$

Answer ___$16\frac{1}{2}$ ft___

2. In the **scalene** triangle MNO, MO = 1.5 ft, MN = 5.5 ft, and NO = 6.5 ft. Find the perimeter of triangle MNO.

Answer ___

3. Is an equilateral triangle always isosceles? _____

4. Is an isosceles triangle always equilateral? _____

Draw a picture. Solve.

5. Find the perimeter of an **equilateral** triangle, one side of which is 4 ft 6 in.

Answer ___4 yards 6 ft___

6. How many feet of wire will be required to enclose a triangle-shaped park whose sides are 150 ft, 140 ft, and 200 ft?

Answer ___490___

7. What is the perimeter of an **isosceles** triangle that has two sides 8 ft long and a third side 6 ft long?

Answer ___7 yds 1 ft___

8. Roger needs trim for a triangular-shaped bandana for his costume. The bandana measures 60 centimeters on 2 sides and 90 centimeters on the 3rd side. How much trim does he need?

Answer ___2100 mm___

9. How many meters of wire will be required to enclose a triangular-shaped park? The sides measure 50, 45, and 69 meters.

Answer ___16400 m___

10. What is the perimeter of an **isosceles** triangle that has two sides 8 kilometers long and a third side 6 kilometers long?

Answer ___22000 m___

Formula for the Area of a Triangle

In rectangle ABCD (at the right) the diagonal AC has been drawn. AC divides the rectangle into two equal triangles.

A triangle is a figure with three sides. The perimeter of a triangle is, of course, the sum of its three sides. We do not speak of a triangle's length and width; we speak of a triangle's **height** and **base.** The **height** of a triangle is the **perpendicular** distance between the base and the opposite vertex (point where two sides meet). In triangle **XYZ** below, the height from the vertex Y to the base XZ is shown by the perpendicular line.

The area of rectangle ABCD is equal to lw, or bh if we use b for base and h for height.

Both triangles in the rectangle are equal, since the diagonal has divided the rectangle into two equal triangles. Therefore, the area of either triangle is $\frac{1}{2} bh$ or $\frac{bh}{2}$.

The area of any triangle is: $A = \frac{1}{2} bh$ or $\frac{bh}{2}$.

EXAMPLE: Find the area of triangle XYZ if the base is 15 feet and the height is 12 feet.

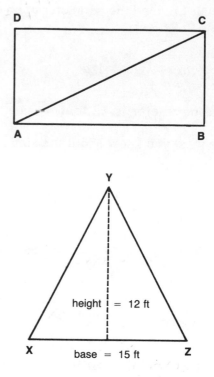

$A = \frac{1}{2} bh$

$A = \frac{1}{2} \times 15 \times 12$

$A = \frac{180}{2} = 90$ The area of XYZ is 90 square feet.

Draw a picture. Find the area of the triangles in these problems.

---

1. What is the area of a triangle whose height is 20 ft and whose base is 16 ft?

    Answer ___160 ft___

2. The railroad cut off a corner of a field on Tim Huarte's farm. The piece cut off formed a triangle whose base was 25 yards and whose height was 30 yards. How many square yards were in the triangle?

    Answer ___375 yrds___

---

3. A park that was the shape of a rectangle was cut into two equal triangles by a fence. The park measured 360 ft by 240 ft. What is the area of each triangle? (See whether you can work this two ways.)

    Answer ___43,200 ft___

4. The gable of a house forms a triangle. If a gable is 10 ft high (height) and the house is 30 ft wide (base), how many square feet are there in the gable?

    Answer ___150 ft___

Right Triangles and Angles

Triangle ABC is a **right triangle** because the two shorter sides (commonly called **legs**) are **perpendicular;** that is, they form a right angle. A right angle contains exactly 90 degrees. Any triangle which has its legs perpendicular is called a right triangle. The side opposite the right angle is the longest side and is called the **hypotenuse.**

If you measure the angles of any triangle, you will learn an important fact: **The sum of the three angles of any triangle is 180 degrees.**

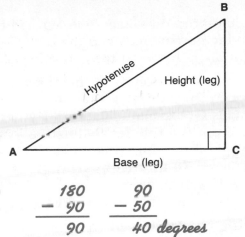

EXAMPLE: Since the sum of the three angles in any triangle is 180 degrees, what is the size of the third angle of a right triangle if the second angle contains 50 degrees?
(180 − 90 − 50 leaves how many degrees?)

$$180 \quad\quad 90$$
$$\underline{-\ 90} \quad \underline{-\ 50}$$
$$90 \quad\quad 40\ degrees$$

Draw a picture. Solve.

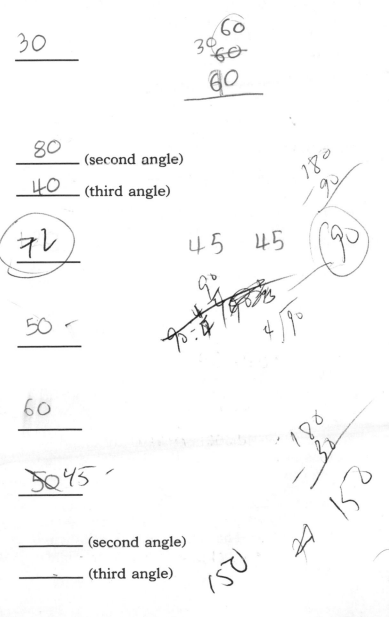

:::::::::::: **Do Your Work Here** ::::::::::::

1. In another triangle (not a right triangle), one angle is 120 degrees, the other two angles are equal to each other. What is the size of each of these two angles?

    30

2. In a certain triangle, one angle is 60 degrees. The second angle is twice the size of the third angle. Find the size of the second and third angles. (First, 180 − 60 = ? Divide this amount into three equal parts—two parts for the second angle, one part for the third angle.)

    80 (second angle)
    40 (third angle)

3. In a right triangle, one angle is 90 degrees. The second angle is four times the third angle. Find the size of the second angle.

    72

4. An isosceles triangle has two equal sides and two equal angles. If each equal angle contains 50 degrees, what is the size of the third angle?

    50

5. An equilateral triangle has three equal sides and three equal angles. Remembering that a triangle contains 180 degrees, what is the size of each angle?

    60

6. In a certain right triangle, the other two angles are equal. How many degrees are there in each of the two equal angles?

    50 45

7. The first angle of a certain triangle contains 30 degrees. The second angle is one half the size of the third angle. How large are the second and the third angles?

    _____ (second angle)
    _____ (third angle)

51

The Pythagorean Theorem

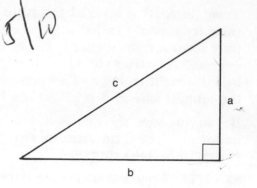

On the preceding page you learned that a right triangle has one leg perpendicular to the other leg. The side opposite the right angle between these two legs is called the hypotenuse.

Now we come to one of the oldest rules about right triangles— **The Pythagorean Theorem,** which was named for the famous Greek mathematician Pythagoras. The Pythagorean Theorem is a rule for finding the length of one side of a right triangle if the other two sides are known. Simply stated, this rule says: **In any right triangle, the square of one leg plus the square of the other leg equals the square of the hypotenuse.**

Using the letters of the triangle above, this rule may be written as $c^2 = a^2 + b^2$. In this rule c^2 means $c \times c$ and is read **c squared.** The square of 4 is 4×4, or 16.

The **square root** of a number is one of the two equal factors of that number. Letters or numbers multiplied together are called factors. Thus the square root of 16 is 4, since $4 \times 4 = 16$. Do you see that the square root of 25 is 5, and the square root of 49 is 7?

$$c^2 = a^2 + b^2$$

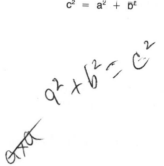

Answer these questions about squares and square roots.

1. What is the square root of 64? __8__   What is the square of 64? __4,096__

2. What is the square root of 81? __9__   What is the square of 81? __6,561__

3. What is the square root of 100? __10__   What is the square of 100? __10000__

4. What is the square root of 144? __12__   What is the square of 12? __144__

5. What is the square root of 400? __20__   What is the square of 20? __400__

Draw a picture. Solve.

▓▓▓▓▓▓ **Do Your Work Here** ▓▓▓▓▓▓

6. If one leg of a right triangle is 3 and the other leg is 4, what is the square of the hypotenuse?

   __25__

7. What is the length of the hypotenuse in problem 6? (Hint: Find the square root of the answer to problem 6.)

   ~~625~~ 5

8. If the legs of a right triangle are 6 and 8 feet long, what is the length of the hypotenuse?

   ~~100~~ 10

9. What is the length of the diagonal of a rectangle whose dimensions are 12 by 16 feet? (Hint: The diagonal of a rectangle forms the hypotenuse of two equal right triangles.)

   ~~400~~ 20

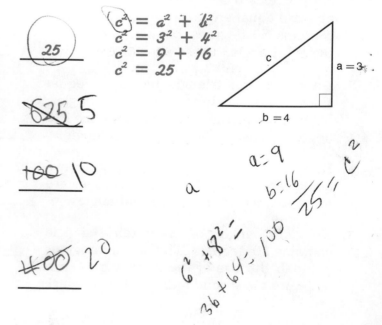

$$c^2 = a^2 + b^2$$
$$c^2 = 3^2 + 4^2$$
$$c^2 = 9 + 16$$
$$c^2 = 25$$

Formula for Circumference of a Circle

A circle is more than a closed curved line. A circle is a curved line, and all points on this curved line are the same distance from a single point called the center of the circle. The distance around a circle is usually called the **circumference.** Every point on the circumference is the same distance from the circle's center.

A straight line drawn across a circle and through its center is called the **diameter** of the circle.

A straight line **from** the center **to** the circumference is the **radius.** The radius is equal to **one half the diameter.** No matter how many radii (plural of radius) are drawn in a circle, they are all of equal length. The same is true of the diameter.

The circumference of a circle is about $\boxed{3.1416 \text{ or } \frac{22}{7}}$ times the diameter. The Greeks used the letter π (called pi) to represent this amount.

The formula for the circumference of a circle is given as $C = \pi d$, or $C = 2\pi r$.

$C = \pi d$

EXAMPLE: Find the circumference of a circle whose diameter is 12 mm.

$C = \pi d$
$C = 3.1416 \times 12$
$C = 37.6992$ Round to 37.7.

The circumference of the circle is 37.7 millimeters.

Solve.

1. What is the circumference of a circle whose diameter is 7 meters? (Hint: Try using $\frac{22}{7}$ for π.)

   Answer _____22_____

2. A circular racetrack at the fairgrounds has a diameter of 175 feet. How long is the track? (The length of the track will be the circumference of the circle.)

   Answer _____549.5_____

3. What is the circumference of a circle whose radius is 7 millimeters? (The radius is $\frac{1}{2}$ the diameter.)

   Answer _____43.96_____

4. The zoo has a circular pool for its polar bears. The pool's diameter is 28 ft. How much fence will be needed to enclose the pool?

   Answer _____87.92_____

5. The wheel of a bicycle has a radius of 35 centimeters. What is the circumference of the wheel?

   Answer _____219.81_____

6. At the Alamo in San Antonio a circular flower garden encloses the star of Texas. The garden has a radius of 3.15 meters. How much fence is needed to enclose the garden? Round your answer to the nearest hundredth if using $\pi = 3.1416$.

   Answer _____19.782_____

53

Formula for Area of a Circle

The formula for finding the area of a circle is $A = \pi r^2$.

The formula, $A = \pi r^2$, means the area of a circle is equal to pi times the radius squared, or the radius times the radius. The radius is $\frac{1}{2}$ the diameter. Use either 3.1416 or $\frac{22}{7}$ for π.

EXAMPLE: Find the area of a circle with a radius of 7 cm.

$A = \pi r^2$

$A = \frac{22}{7} (7 \times 7)$

$A = \frac{22}{7} \times 49 = 154$

radius = 7 cm

The area of the circle is 154 square centimeters.

Remember, write your answer in *square* units.

Use the formula for area of a circle. Solve.

---

1. Find the area of a circle that has a radius of 3.5 meters. Round your answer to the nearest tenth.

Answer _____38.465_____

2. What is the area of a circle that has a diameter of 28 meters? (Hint: To change to radius, divide the diameter by 2.) Round your answer to the nearest one.

Answer _____615.44_____

---

3. A bandstand in the shape of a circle is to be 7 meters across. How many square meters of flooring will be required? (Hint: The radius is one half the diameter.) Round your answer to the nearest hundredth if needed.

Answer _____38.465_____

4. The circular canvas net used by fire fighters has a radius of 2.1 meters. What is the area of the net? Round your answer to the nearest tenth.

Answer _____13.8000_____

---

5. A gallon of paint covers 56 square meters. How many gallons of paint will be needed to cover a circular floor that is 22 meters in diameter? (Hint: the radius is one half the diameter.) Round your answer to the nearest one.

Answer _____400_____

6. The top of a piston is a circle. The top of an auto piston has a radius of 4.2 centimeters. What is the area of the top of the piston? Round your answer to the nearest hundredth if needed.

Answer _____55.40_____

54

Using Formulas with Circle Problems

For finding the circumference, use the formula $C = \pi d$. For finding the area, use the formula $A = \pi r^2$. Use $\pi = \frac{22}{7}$ in these problems.

EXAMPLE: Gloria has an electric train which runs on a circular track 14 feet in diameter. How many times will the train go around in traveling 1 mile?

$C = \pi d$

$C = \frac{22}{7} \times 14$

$C = 44$ feet around

To find how many 44-foot laps in a mile, divide.

1 mile = 5,280 ft

$\frac{5,280}{44} = 120$ laps in a mile

$d = 14$ ft

Solve these problems dealing with circles.

1. The wheel of Dan's bicycle has a diameter of 28 inches. How many revolutions will it make in going one mile? Convert your answer to feet. (Hint: First find the circumference of the wheel.)

   Answer _____720_____

2. The inside diameter of the ring of a basketball goal is 18 in. Find the length of the circumference of that ring. Round your answer to the nearest hundredth.

   Answer ____56.50____

3. Jerry tied his horse to a stake with a rope 15 meters long. Allowing 1 meter for tying the knot, over how many square meters can the horse graze?

   Answer _____616_____

4. The top of an auto piston is round in shape. What is the surface area of the top of a piston whose diameter is $3\frac{1}{2}$ inches?

   Answer ___10.99   9.62___

5. There are 360 degrees in a circle. If a clock shows exactly 4:00 o'clock, how many degrees are there in each angle formed by the hands of the clock? (Since 4 hours are $\frac{1}{3}$ of 12 hours, $\frac{1}{3}$ of 360 is how many degrees? Find the other angle by subtraction.)

   Answer ___120   240___

6. A circular flower garden, 32 feet in diameter, is surrounded by a concrete walk which is 2 feet wide. Find the area of the walk. (Subtract the area of the inner circle from that of the outer circle: $\pi R^2 - \pi r^2$.) Round to the nearest one.

   Answer _____

Remembering that r is $\frac{1}{2}d$ and using the formulas, fill in the blanks.

7.

r	d	C	A
14			
	14		
7			

8.

r	d	C	A
	42		
3.5			
	7		

9.

r	d	C	A
21			
	21		
	3.5		

The Cylinder

This figure is called a **cylinder.** You have seen many cylinders, such as cans, tanks, pipes, and the like. A cylinder has two bases, which are parallel and equal. A cylinder differs from a circle in that it has **height** (or **altitude**). Since the bases of a cylinder are equal and parallel, the height is the perpendicular distance between the bases. The height will be the same whether drawn in the center, as shown in the illustration, or measured along the edge.

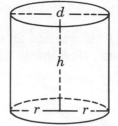

One person's definition of a cylinder was "a stack of equal circles." This definition gives a good clue to finding the volume of a cylinder. We know how to find the area of a circle $(A = \pi r^2)$. If we multiply this base area (the base is a circle) by the height, we shall have the volume of the cylinder. Therefore, **the formula for finding the volume of any cylinder is** $V = \pi r^2 h$. Using this formula, find the volume of the cylinders in the problems below.

EXAMPLE: Find the volume of a cylinder that has a radius of 7 feet and a height of 10 feet.

$$V = \pi r^2 h$$

$$V = \frac{22}{7} \times 7 \times 7 \times 10$$

$$V = 1,540 \text{ cubic feet}$$

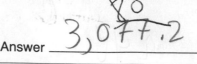

$r = 7$ ft

$h = 10$ ft

Solve these volume problems. Use $\pi = \frac{22}{7}$.

---

1. A water standpipe is 60 feet high and has a diameter of 14 feet. What is the volume of the standpipe (in cubic feet)?

Answer ___9,231.6___

2. A water tank the shape of a cylinder has a diameter of 15 ft and is 35 ft high. How many cubic feet of water will it hold?

Answer ___6,181.875___

---

3. A gasoline storage tank has a diameter of 28 ft and a height of 30 ft. If 1 cu ft holds about 7.5 gallons, how many gallons will the tank hold?

Answer ___2,461.76 gallons___

4. A tank with a diameter of 14 ft is filled to a depth of 20 ft with water. How many cubic feet of water does it contain?

Answer ___3,077.2___

---

5. Two cylinders have the same height—20 ft. One has a 7 ft radius; the other has a radius of 14 ft. The volume of the second cylinder is how many times that of the first? (Do not guess, or you will be fooled. Doubling the radius increases the volume how many times?)

Answer ___9,41.6___

6. A can in the shape of a cylinder is 6 in. wide (diameter) and 12 in. high. What is the volume, or capacity, of the can in cubic inches?

Answer ___339.12___

Using Mixed Units of Measurement to Find Perimeter

When dimensions are not given in the same unit of measure, we must change one measure to agree with the other. For example, if the length is given in feet and the width is given in inches, both must be changed to feet or both must be changed to inches before they can be added.

EXAMPLE: A desk measures 5 feet by 40 inches. What is its perimeter?

$l = 5$ feet $\quad w = 40$ inches

Convert measures to the same units. Change inches to feet.

12 inches = 1 foot

$\frac{40}{12} = 3\frac{1}{3}$ feet

(Changing the units to inches is also correct.)

$$P = 2l + 2w$$
$$P = 2(5) + 2\left(3\frac{1}{3}\right)$$
$$P = 10 + 2\left(\frac{10}{3}\right)$$
$$P = 10 + \frac{20}{3}$$
$$P = 10 + 6\frac{2}{3}$$
$$P = 16\frac{2}{3} \text{ feet}$$

40 in.

5 ft

Solve.

1. Find the perimeter of a rectangle whose length is 14 yards and whose width is 40 ft.

   Answer _____

2. How many feet of molding will be required to go around the floor of a room measuring 24 ft by 31 ft 6 in.?

   Answer _____

3. Janet's farm is a mile long and lacks 80 yards of being a mile in width. Express the perimeter in yards. (1,760 yards = 1 mile)

   Answer _____

4. Lions Field, a playground donated by the Lions Luncheon Club, is $\frac{1}{4}$ mile long and 528 feet deep (wide). Express the perimeter in miles or feet.

   Answer _____

5. How many feet of fence will be needed to enclose the play field measuring 70 yd by 250 ft?

   Answer _____

6. A park measuring 250 yd by 400 ft has a sidewalk around it. What is the least distance a person must walk to go around the park once?

   Answer _____

7. Edith bought decorative trim to put around a blanket that measured 39 in. by 42 in. How much did she pay for the trim at 80 cents a yard?

   Answer _____

8. What is the perimeter of a rectangle that is twice as long as it is wide? It is 20 ft 6 in. wide.

   Answer _____

Using Mixed Units of Measurement to Find Area

Study the table of square measures on the right to answer these problems.

EXAMPLE: Find how many square inches are in 2 square feet.

> 1 sq ft = 144 sq in.
>
> $2 \text{ sq ft} = 2 \times 144 = 288 \text{ sq in.}$

> 1 sq ft = 12 × 12 = 144 sq in.
> 1 sq yd = 3 × 3 = 9 sq ft
> 1 acre (A) = 4,840 sq yd
> 1 sq mi = 640 A

1. To change square inches to square feet, divide by _____.

2. 432 square inches equal how many square feet? _____

3. To change square feet to square yards, divide by _____.

4. 27 square feet equal how many square yards? _____

5. To change square yards to square feet, multiply by _____.

6. Four square yards equal how many square feet? _____

7. To change square yards to acres, divide by _____.

8. 121 square yards equal how many acres? _____

9. A library floor has the dimensions of 70 ft by 30 ft. How many square yards are there in the floor area? _____

Solve these word problems.   Do Your Work Here

10. Find the cost, at $4.50 per square yard, of laying a concrete walk 100 yd long and $1\frac{1}{2}$ yd wide. (Find the area, then the cost.) _____

11. We put tile on the kitchen floor at a cost of $10.50 per sq yd. The floor measures 18 ft by 14 ft. What was the cost? _____

12. Estelle had the living room rug cleaned at 40¢ a square foot. The rug measures 9 ft by 12 ft. What did the cleaning cost? _____

13. The living room (*problem 12*) where the 9 by 12 rug is placed measures 13 ft by 16 ft. How many square feet of floor are not covered by the rug? (Find the total floor area, then subtract the rug area.) _____

14. We bought new carpet at a cost of $10 per square yd. The floor measures 54 ft by 42 ft. What was the cost? _____

Using Mixed Units of Measurement to Find Volume

Use the table at the right in solving problems with volumes.

1 cu ft = 1,728 cu in. (12 × 12 × 12) 1 cu yd = 27 cu ft (3 × 3 × 3)

EXAMPLE: How many bushels of corn can be stored in a crib measuring 15 ft by 6 ft by 10 ft? (Count 1.25 cu ft to the bushel.)

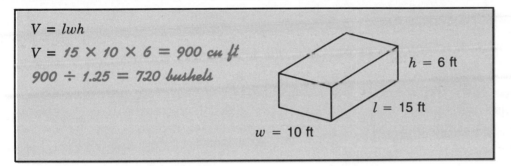

$V = lwh$

$V = 15 \times 10 \times 6 = 900$ cu ft

$900 \div 1.25 = 720$ bushels

$h = 6$ ft
$l = 15$ ft
$w = 10$ ft

Solve these volume problems.

1. Mr. Harper plans to lay fresh soil 4 in. deep on his yard. The yard is 60 ft by 54 ft. How many cubic yards of soil does he need?

    Answer _____

2. Pat Harbin bought a pile of wood 20 ft long, 16 ft wide, and 8 ft high. How many cords did she buy? (1 cord = 128 cu ft)

    Answer _____

3. How much did she (*problem 2*) pay for the wood at $15.50 per cord?

    Answer _____

4. A cubic foot of water weighs 62.5 lb. How much does the water weigh in a filled tank that is 10 ft by 12 ft by 8 ft?

    Answer _____

5. In excavating a hole 80 ft long, 54 ft wide, and 10 ft deep, how many loads of dirt had to be moved in trucks that hold 5 cubic yards?

    Answer _____

6. How many cubic yards of concrete will Ms. Chung have to buy to build a driveway 6 ft wide, 108 ft long, and 4 in. thick?

    Answer _____

7. There are 231 cu in. to a gallon. The gas tank of Henry's car measures 33 in. by 14 in. by 12 in. How many gallons will it hold?

    Answer _____

8. What is the volume of a rectangular storage tank that measures 24 ft by 22 ft by 11 ft?

    Answer _____

9. What would be the cost of excavating a basement to measure 54 ft by 40 ft by 12 ft at $7 per cubic yard?

    Answer _____

Using Mixed Units of Measurement to Find Volume

The volume of any container may be expressed in cubic feet or inches. However, the volume of some containers is commonly expressed in units like gallons or bushels. Using the equivalents listed below for changing cubic units to gallons or bushels, solve these problems.

231 cu in. = 1 gallon	$7\frac{1}{2}$ gallons = 1 cu ft
2,150.42 cu in. = 1 bu	1.25 cu ft = 1 bu

Solve.

1. Gil's Feed Store has a bin which is 20 ft long and 10 ft wide and is filled with grain to a depth of 6 ft. How many bushels does it hold?

   Answer _____

2. A storage bin at the grain elevator is 60 ft by 40 ft and is filled to a depth of 20 ft. How many bushels does it hold?

   Answer _____

3. The swimming pool is 45 ft long and 30 ft wide. If it is filled to an average depth of 6 ft, how many gallons of water does it hold?

   Answer _____

4. Our car has a gas tank which measures 42 in. by 11 in. by 10 in. How many gallons will it hold?

   Answer _____

5. A coal car has inside dimensions of 40 ft by 6 ft and is filled level to a 5-ft depth. Allowing 40 cu ft to the ton, how many tons of coal are there in this car?

   Answer _____

6. How many gallons will a can hold if it measures 11 in. by 10 in. by 10.5 in.?

   Answer _____

7. How many bushels can be stored in a bin 8 ft wide and 10 ft long if it is filled to a depth of 5 ft?

   Answer _____

8. How many gallons will a can that measures $5\frac{1}{2}$ in. × 5 in. × 21 in. hold?

   Answer _____

Finding the Missing Factors

When multiplying two or more numbers, each number is called a **factor** of the product. Thus, in $6 \times 8 = 48$, 6 and 8 are factors of 48. What is the other factor of 24 when 3 is one of the factors? Study the examples below.

Find: $3 \times$ _____ $= 24$

> Think: $24 \div 3 = ?$
> Since $24 \div 3 = 8$, $3 \times 8 = 24$.
> 3 and 8 are factors of 24.

Find: $60 \div$ _____ $= 6$

> Think: $6 \times ? = 60$
> Since $6 \times 10 = 60$, $60 \div 10 = 6$.
> 6 and 10 are factors of 60.

Find the missing factors.

1. __8__ $\times 4 = 32$
   (**Think:** $32 \div 4 = ?$)

2. $6 \times$ __5__ $= 30$

3. __8__ $\times 7 = 42$

4. __5__ $\times 8 = 40$

5. $3 \times$ __8__ $= 24$

6. $9 \times$ __4__ $= 36$

7. $10 \times$ __15__ $= 150$

8. __15__ $\times 8 = 120$

9. $32 \times$ __3__ $= 96$

10. $35 \times$ __2__ $= 70$

11. $50 \div 2 =$ __25__

12. $50 \div 25 =$ __2__

13. $15 \div$ __5__ $= 3$
    (**Think:** $3 \times ? = 15$)

14. $15 \div$ __3__ $= 5$

$48 \div$ __12__ $= 4$

$36 \div$ __12__ $= 3$

$28 \div$ __7__ $= 4$

$75 \div$ __3__ $= 25$

$75 \div$ __25__ $= 3$

$120 \div$ __12__ $= 10$

$120 \div$ __10__ $= 12$

$65 \div$ __13__ $= 5$

__24__ $\div 4 = 6$
(**Think:** $4 \times 6 = ?$)

__60__ $\div 12 = 5$

__44__ $\div 11 = 4$

__56__ $\div 8 = 7$

__100__ $\div 20 = 5$

__75__ $\div 15 = 5$

__45__ $\div 15 = 3$

__90__ $\div 15 = 6$

__60__ $\div 12 = 5$

__40__ $\div 20 = 2$

__96__ $\div 16 = 6$

__84__ $\div 14 = 6$

__72__ $\div 8 = 9$

__84__ $\div 7 = 12$

__91__ $\div 13 = 7$

__180__ $\div 30 = 4$

__600__ $\div 50 = 12$

__11__ $\div 5.5 = 2$

__12__ $\div 1.2 = 10$

__125__ $\div 12.5 = 10$

61

Finding the Missing Factor in the Interest Formula

When the product and one factor are given, the other factor can be found. Thus, $7 \times 5 = 35$, and $35 \div 7 = 5$, and $35 \div 5 = 7$. In a similar manner, **any one factor may be found if the product and all other factors are known.**

$12 \times 10 \times 20 = 2{,}400$, and $\frac{2{,}400}{10 \times 20} = 12$, and $\frac{2{,}400}{12 \times 20} = 10$, and $\frac{2{,}400}{12 \times 10} = 20$

Apply this skill to find missing factors in the interest formula, $I = prt$. Then, $p = \frac{I}{rt}$ and $r = \frac{I}{pt}$ and $t = \frac{I}{pr}$.

The rule for finding a missing factor is: **divide the interest by the product of the known factors.**

EXAMPLE: If the interest on $200 for 3 years is $24, what is the rate of interest?

$I = prt$

$\$24 = \$200 \times r \times 3$ The rate is the missing factor.

$r = \frac{I}{pt}$

$r = \frac{\$24}{\$200 \times 3} = \frac{\$24.00}{\$600} = 0.04 = 4\%$

Solve each problem below.

1. Ted paid $12 interest on a loan of $100 for 1 year. What rate of interest did he pay?

Answer ___12%___

2. When I is $15 and p is $300 and t is 1, what rate of interest was this?

Answer ___5%___

3. Ms. Black paid $32 interest at the rate of 8% on a loan of $200. For how long was the loan? (Hint: Change 8% to a decimal.) $\left(t = \frac{I}{pr}\right)$

Answer ___2 yrs___

4. Interest of $120 at 12% was paid on a loan of $500. For how long was the loan?

Answer ___2 yrs___

5. Find t when I is $12.50, r is 5%, and p is $125.

Answer ___2 yrs___

6. Interest of $24 was paid on a 6% loan for 2 years. What was the amount of the loan? $\left(p = \frac{I}{rt}\right)$

Answer ___$200___

7. Joe earned $5 in interest on his savings account in 1 year. The interest rate was 5% simple interest, not compound. How much did he have in the savings account?

Answer ___$100___

8. How much did Arturo have in the bank at 5% simple interest if the interest amounted to $12 in 2 years?

Answer ___$120___

Finding the Missing Factors
in the Area Formula

If you know the area and either the length or width of a rectangle, you can find the missing dimension. If $A = lw$, then $l = \frac{A}{w}$ and $w = \frac{A}{l}$. For example, if $8 \times 4 = 32$, then $\frac{32}{4} = 8$ and $\frac{32}{8} = 4$.

EXAMPLE: Find the width of a rectangle with length 24 feet and area 480 square feet.

$A = lw$
$480 = 24 \times w$ w is the missing factor.
$w = \frac{A}{l}$
$w = \frac{480}{24} = 20$ The width is 20 feet.

$A = 480$ sq ft $w = ?$
$l = 24$ ft

Solve each word problem.

1. The length of a rectangle is 28 meters and its area is 560 square meters. Find the width.

   Answer ___20___

2. One field on Jack's ranch contains 20,000 square yards. The length of the field is 200 yards. What is the width of this field?

   Answer ___100___

3. The lot on which Nancy built her new house contains 7,500 square feet. The lot is 150 feet deep (long). How wide is it?

   Answer ___50___

4. A rectangle has an area of 30,000 square feet and has a width of 200 feet. What is the length of this rectangle? ($l = \frac{A}{w}$)

   Answer ___150___

5. Another rectangle has an area of 240 sq in. and has a width of 12 in. What is the length of this rectangle?

   Answer ___20___

6. The land on which the school is built contains 200,000 square feet. It is 400 feet wide. How long is it?

   Answer ___500___

7. Gina's house is built on a lot containing 11,250 square feet and is 75 feet wide. What is the length of this lot?

   Answer ___150___

8. The courthouse is built on a piece of land containing 30,000 square feet and having a width of 150 feet. What is its length?

   Answer ___200___

Fill in the blanks in this table.

9.	l	20	20	18	25	50					
10.	w						15	30	19	15	25
	A	200	400	360	750	2,500	600	1,200	380	450	625

63

Finding the Missing Factors
in the Volume Formula

The formula for finding the volume of a rectangular prism is V = lwh. From that formula come these: $l = \frac{V}{wh}$, and $w = \frac{V}{lh}$, and $h = \frac{V}{lw}$. Therefore, if the volume and two dimensions of a rectangular prism are known, we can easily find the other dimension.

EXAMPLE: Find the height of a box with length of 6 inches, width of 4 inches, and volume of 240 cubic inches.

$V = lwh$
$240 = 6 \times 4 \times h$ The height is the missing factor.
$h = \frac{V}{lw}$

$h = \frac{240}{6 \times 4} = \frac{240}{24} = 10$ The height is 10 in.

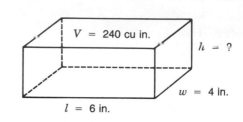

$V = 240$ cu in.
$h = ?$
$w = 4$ in.
$l = 6$ in.

Solve these problems.

1. The volume of a rectangular solid is 480 cubic inches. The length is 12 inches, and the width is 10 inches. What is the height?

   Answer ___4___

2. Frank has a rectangular-shaped gas can whose volume is 1,800 cubic inches. The base of the can measures 12 inches by 10 inches. How deep (high) is the can?

   Answer ___15___

3. Remember that a gallon contains 231 cubic inches. Tim has a can containing exactly a gallon. The base of the can is 7 in. by 3 in. How tall is the can?

   Answer ___11___

4. A rectangular solid has a volume of 1,800 cubic inches. Its height is 15 inches, and the width of the base is 12 inches. What is the length of the base? ($l = \frac{V}{wh}$)

   Answer ___10___

5. A bin has a volume of 960 cubic feet. The bin is 20 feet wide and 8 feet deep. How long is the bin?

   Answer ___6___

6. An open tank, rectangular in shape, has a capacity of 2,520 cubic feet. The tank is 15 feet deep and 14 feet wide. How long is it?

   Answer ___12___

7. A storage tank, rectangular in shape, has a capacity of 15,000 cubic feet. It is 30 feet deep and 25 feet long. How wide is it? ($w = \frac{V}{lh}$)

   Answer ___20___

8. Fill in the missing dimensions.

l	10			7	8	10
w	20	6	10	8		
h	30	12	15		12	12
V	6,000	720	750	560	960	480

Using the Distance Formula

The distance formula shows the relationship between time, rate of speed, and distance traveled. $d = rt$ means that distance equals the rate times the time.

EXAMPLE 1: If we averaged 45 miles per hour for 4 hours, how far did we travel?

$d = rt$
$d = 45 \times 4$
$d = 180$ miles

EXAMPLE 2: The distance from Dallas to New York by road is 1,648 miles. A team of drivers drove this distance and averaged 41.2 miles per hour. How many hours did the trip take?

$d = rt$
$1,648 = 41.2 \times t$ The missing factor is time.
$t = \frac{d}{r}$
$t = \frac{1,648}{41.2} = 40$ hours

EXAMPLE 3: An airplane flew 500 miles in 1 hr and 15 min. What was its speed, or rate? (Change 15 min to 0.25 hr)

$d = rt$
$500 = r \times 1.25$ The missing factor is rate.
$r = \frac{d}{t}$
$r = \frac{500}{1.25} = 400$ miles per hour

1.06

Solve the problems below.

1. A passenger plane makes the flight from Milwaukee to Detroit, a distance of 330 miles, in 1 hour and 6 minutes. What is its speed?

Answer ___300___

2. From Chicago to New York via one airline is 900 miles. The flight is made in 2 hours and 30 minutes. What is this speed?

Answer ___360___

3. An airplane flew the 630 miles from Cleveland to Memphis and averaged 360 miles per hour. How many hours did the flight take?

Answer ___1.75___

4. The distance from Phoenix to El Paso is 400 miles by road. How many hours are needed to make the trip if you average 50 miles per hour?

Answer ___8___

5. On our vacation we averaged 45 miles per hour. At this rate, how far did we travel in 8 hours and 20 minutes?

Answer ___375___

6. The maximum speed of a certain airplane is about 600 miles per hour. If the plane flew at this speed for 15 minutes, how far would it travel? (15 min $-\frac{1}{4}$ hr)

Answer ___~~56~~ 150___

7. From Louisville, Kentucky to Chicago by road is 300 miles. If we averaged 50 miles per hour, how long did the trip take?

Answer ___6___

8. How far can you travel in 10 hours if your average speed is 43.6 miles per hour?

Answer ___436___

Finding the Interest Rate on Installment Purchases

Buying on the installment plan is a convenience. There are several methods used to compute the annual rate of interest. The formula at the right lets you use one of the more common methods.

The Formula

$$\text{Annual Rate of Interest} = \frac{2 \times \frac{\text{number of payments possible in one year}}{} \times \frac{\text{total interest}}{}}{\text{total loan amount} \times \text{number of payments made plus 1}}$$

An installment purchase is really a loan. In the example below, the **total loan amount** is the difference between the cash price and the down payment, or $75. The **total interest** paid is the difference between the total installment payments ($4 \times \$20 = \$80$) and the amount of the "loan" ($\$80 - \$75 = \$5$). Since Madge will make monthly payments, the **number of payments possible in one year** is 12. (This figure would be 52 if the payments were made weekly.) The **number of payments made plus 1** is 5.

EXAMPLE: Madge bought a $90 set of dishes. She made a down payment of $15 and agreed to make 4 monthly payments of $20 each. What rate of interest did she pay?

$$\text{Annual Rate} = \frac{2 \times 12 \times \$5}{\$75 \times 5} = \frac{\$120}{\$375}$$

Divide: $\$120 \div \$375 = 0.32 \text{ or } 32\%$

The Annual Interest Rate = 32%

Find the rates of interest in these problems.

1. Ana McClure bought a home freezer. It was priced at $400. She paid $200 down and agreed to make 9 monthly payments of $24 each. What rate of interest did she pay?

   Answer _____

2. Bernard bought a small radio priced at $30. He made a down payment of $10 and made 5 monthly payments of $4.25 each. What rate of interest was this?

   Answer _____

3. Bill bought a shotgun priced at $50. He paid $30 down and agreed to pay the balance with 4 monthly payments of $5.50 each. What rate of interest did he pay?

   Answer _____

4. Marie Duval bought an electric sewing machine priced at $150. She paid $50 cash and agreed to make 9 monthly payments of $12 each. What rate of interest was this?

   Answer _____

5. Joe Stevens bought a car for $2,195 and made a cash payment of $395. He agreed to make 24 monthly payments of $90 each. What rate of interest did he pay?

   Answer _____

6. Jim Foxx bought an electric refrigerator priced at $300. He made a cash payment of $150 and agreed to make 7 monthly payments of $24 each. What rate of interest was this?

   Answer _____

The Bar Graph

A graph is a picture of data (factual information). A bar graph shows information in the shape of bars. Use a bar graph to answer questions about the data.

This bar graph shows the number of cups of each flavor of frozen yogurt sold at The Freezer during the summer.

EXAMPLE: Find how many more cups of strawberry were sold than cups of chocolate?

Use these steps when finding the needed data.

1. Locate the flavor on the horizontal scale along the bottom of the graph.

2. For each flavor, move to the top of the bar and back to the vertical scale along the side of the graph.

3. Write down the data.
Data: 300 cups of strawberry were sold.
175 cups of chocolate were sold.

Subtract.
300 − 175 = 125
There were 125 more cups of strawberry sold than of chocolate.

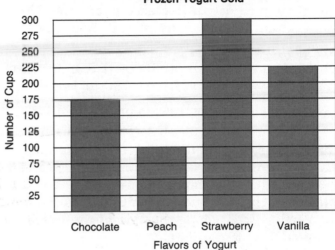

Frozen Yogurt Sold

275
300
250

Solve.

---

1. How many cups of yogurt were sold in all?

Answer _____800_____

2. Which flavors of yogurt had sales of less than two hundred?

Answer __chocolate and peach__

---

3. Which flavor of yogurt had the least number of sales?

Answer _____Peach_____

4. How many more cups of vanilla were sold than cups of peach?

Answer _____125_____

---

5. How many fewer cups of chocolate were sold than cups of vanilla?

Answer _____50_____

6. How many more cups of strawberry were sold than cups of peach?

Answer _____200_____

The Line Graph

A line graph shows how something can change over a period of time. The dots represent data (factual information). Lines connect the dots and show increase or decrease over time. Use a line graph to answer questions about the data.

This graph shows the profits of JBS Industries over a six-year period from 1980 to 1985.

EXAMPLE: Find the increase in profits from 1980–1981.

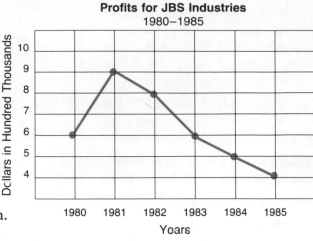

Profits for JBS Industries
1980–1985

Use these steps when finding the needed information.

1. Locate the year on the horizontal scale along the bottom of the graph.
2. For each year, move up the line to the dot and back to the vertical scale along the side of the graph.
3. Write down the data.
   Data: In 1980 JBS Industries earned $600,000. (6 × 100,000)
   In 1981 JBS Industries earned $900,000. (9 × 100,000)

Subtract.
$900,000 − $600,000 = $300,000

The profits for JBS Industries increased by $300,000.

Solve.

1. The line graph shows the profits for JBS Industries for how many years?

   Answer ____6____

2. What is the dollar value of each number at the left of the graph?

   Answer __hundred thousands__

3. What was the decrease in profits between 1981 and 1982?

   Answer __1 hundred thousand__

4. What was the decrease in profits between 1983 and 1984?

   Answer __1 h th__

5. What was the decrease in profits between 1981 and 1985?

   Answer __5 h th__

6. Between which two successive years did the profits decrease the most? How much was the decrease?

   Answer __1982 – 1983 2 h th__

The Circle Graph

A circle graph shows how a whole or total amount or 100% can be separated into parts, percentages or fractions. This circle graph shows how one family plans to use its earnings during the year.

EXAMPLE: Find the amount of money for each part of the family budget. The income for this family is $28,000 per year.

1. *Shelter:* 35% of $28,000 = *$28,000 × 0.35 = $9,800*

Clothing: 15% of $28,000 = ______

Food: 25% of $28,000 = ______

Misc. Exp.: 20% of $28,000 = ______

Ins. and Savings: 5% of $28,000 = ______

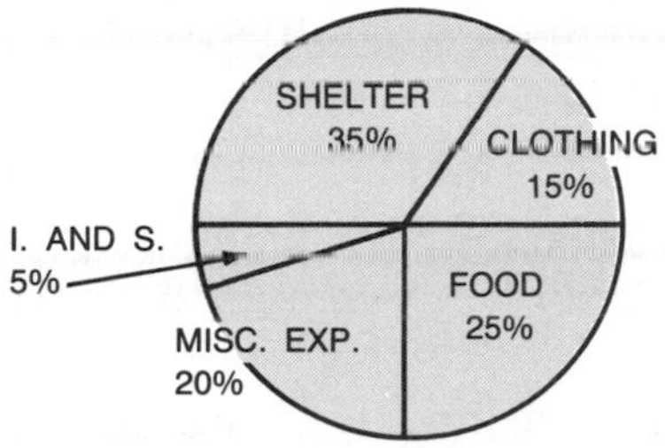

To check, add all five amounts. The total should be $28,000.

2. What percent of the budget is devoted to both food and shelter? How much is actually spent on food and shelter together? ______

3. What total percent was spent on clothing, food, and shelter? ______

4. How many more dollars were spent on food than on clothing? ______

The circle graph at the right shows how one city divides its expenditures among the various agencies of the city.

BREAKDOWN ON GENERAL FUND
(City Operating Budget-$12,000,000)

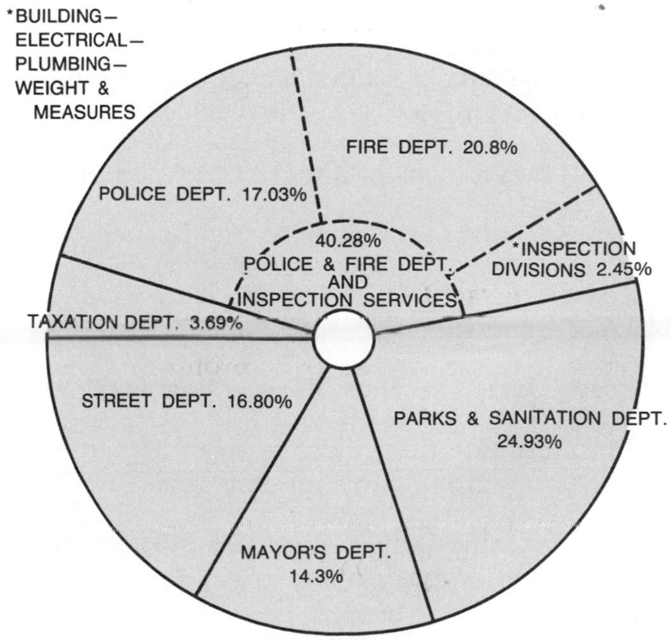

5. Which department spent the most?

______

6. Which department spent the least?

______

7. How much money was spent by the street department? ______

8. Police and fire departments are usually administered as one unit. What percent did this unit (the two departments) spend?

______

9. How much was spent by the mayor's department? ______

Reading Meters and Tables

Most homes have electric meters with four dials like the ones shown below. These dials move to show the usage of electricity in kilowatt-hours.

To read the meter, begin at the leftmost dial. Record the smaller of the two numbers that the hand on the dial is between. Do this for all the dials. The four numbers show the current reading of kilowatt-hours (KWH) used.

1428 kilowatt-hours

Gas meters have three dials that are read in the same way as electric meters. Gas usage is measured in units of a hundred cubic feet, or CCF.

EXAMPLE: The meter above shows a customer's reading for August 1. If the reading for July 1 was 0637, how many kilowatt-hours were used during July? If the cost of electricity is $0.07 per KWH, how much will the customer be charged?

Subtract to find the number of KWH used. $1428 - 0637 = 791$
Multiply 791 by the cost per KWH. $791 \times \$0.07 = \$55.37$

Write the meter reading.   **Solve.**

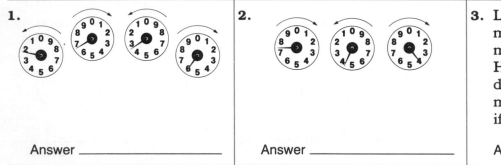

1.

Answer _____

2.

Answer _____

3. Last month Bertha's electric meter read 1574 KWH. This month it reads 1838 KWH. How many kilowatt-hours did she use during the month? How much is the bill if the cost is $0.08 per KWH?

Answer _____

To find the mileage between two cities from a table, find one of the cities in the left-hand column. Lay a ruler along that row to the column headed by the other city's name. Find the road distance from Atlanta to New York.

EXAMPLE: Atlanta is the first city in the left column. Move along that row to the column for New York. The distance is 868 miles.

4. How far is it from Atlanta to Los Angeles? _____

5. How far is it from Boston to San Francisco? _____

6. From Chicago to Los Angeles is how many miles by road? _____

7. From Detroit to New York is how far? _____

8. The Fontaine family drove from Montreal to Miami. How many miles is this? _____

9. The Fontaine family averaged 30 miles to the gallon of gas on the trip. How many gallons did they use? _____

AUTOMOBILE MILEAGE	Chicago, Ill.	Denver, Colo.	Houston, Tex.	Los Angeles, Calif.	Miami, Fla.	New York, N.Y.	San Francisco, Calif.
Atlanta, Ga.	671	1436	852	2245	663	868	2579
Boston, Mass.	992	2016	1865	3004	1615	220	3265
Chicago, Ill.		1062	1139	2115	1352	824	2240
Cleveland, Ohio	311	1393	1372	2393	1327	493	2571
Dallas, Tex.	923	820	241	1476	1367	1580	1790
Denver, Colo.	1062		1061	1148	2104	1794	1324
Detroit, Mich.	271	1333	1306	2415	1437	670	2511
Houston, Tex.	1139	1061		1585	1256	1714	1950
Las Vegas, Nev.	1905	843	1426	305	2572	2637	624
Los Angeles, Calif.	2115	1148	1585		2841	2784	411
Memphis, Tenn.	579	1054	560	1816	1050	1114	2214
Miami, Fla.	1352	2104	1256	2841		1395	3192
Montreal, Quebec	804	1866	1839	2948	1749	381	3044

Complete the charts.

1. Find the perimeter and area of the rectangles.

l	w	P	A
12	8	40	96
15	12	54	180
22	20	84	440

2. Find the volume of the rectangular solids.

l	w	h	V
12	8	10	960
15	10	20	3,000
24	20	10	4,800

3. Find the area of the triangles.

a	b	A
24	18	216
30	20	300
44	15	330

4. Find the circumference of the circles. Use $\pi = \frac{22}{7}$.

r	d	C
7	14	43.96
7	14	43.96
10.5	21	65.94

5. Find the area of the circles. Use $\pi = \frac{22}{7}$.

r	d	A
14	28	615.44
3.5	7	38.465
21	42	1,384.74

6. Find the volume of the cylinders. Use $\pi = \frac{22}{7}$.

r	d	h	V
7	14	10	1,538.6
21	42	12	16,616.88
14	28	20	12,308.8

Change each measurement.

7. 3 mi = _5,280_ yd

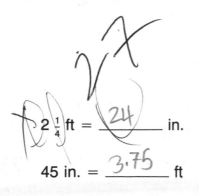

$2\frac{1}{4}$ ft = _24_ in.

9.2 L = _9,000.2_ mL

8. 12 qt = _6_ pt

45 in. = _3.75_ ft

4,139 m = _4,139000_ km

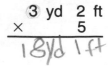

Find each answer.

9.
```
   8 lb   9 oz
 + 1 lb  12 oz
```
10 lb 5 oz

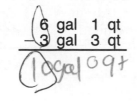
```
   6 gal  1 qt
 - 3 gal  3 qt
```
10 gal 0 qt

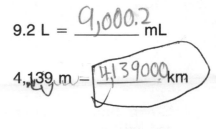
```
   3 yd  2 ft
 ×         5
```
18 yd 1 ft

Problem-Solving Strategy: Make a Diagram

When you read a problem, a diagram may help you decide what to do. A rough sketch is usually good enough. Make sure to label the drawing with the facts given in the problem.

STEPS
1. Read the problem.

Juanita wants to tile the floor of her workshop. The floor measures 13 feet long and 10 feet wide. How many square feet of tile will she need?

2. Make a diagram.

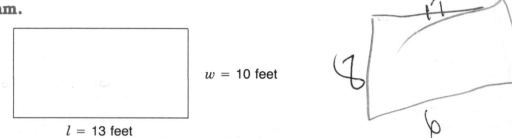

$w = 10$ feet

$l = 13$ feet

3. Solve the problem.

The floor of the workshop is in the shape of a rectangle. To know how many square feet of tile Juanita will need, find the area of the rectangle. Use the formula $A = lw$.

$$A = l \times w$$
$$A = 13 \times 10$$
$$A = 130$$

The area of the workshop floor is 130 square feet. Juanita will need 130 square feet of tile.

Solve.

1. Andrew Willis has a garden 8 meters long and 6 meters wide. If a gate is 0.5 meter wide, how many meters of fence will he need to enclose the garden?

   Answer _____27.5_____

2. Mary Mathews wishes to put a strip of trim, called molding, around the bottom and the top of her living room walls. The room is 16 feet by 14 feet. How many feet of molding are needed?

   Answer _____2\120_____

3. Kevin Sullivan wished to have his yard sodded. The yard is 60 ft by 40 ft. About how many square feet of sod does he need?

   Answer _____2,400_____

4. Andrea bought some broadloom carpeting. She bought a piece measuring 9 feet by 15 feet. How many square feet did she buy?

   Answer _____135._____

Applying Your Skills

Make a diagram to help solve the following problems.
Use all your facts.

Solve.

1. Aaron Jacobi had sheet vinyl installed at $11.50 per square yard. The floor measures 36 feet by 50 feet. How much did the vinyl cost?

   Answer __69.00__

2. The school gymnasium floor is 54 feet by 100 feet. What was the cost of sanding the floor at $1.25 per square yard?

   Answer __22.50__

3. Mr. Brown is having some painting done. The area to be painted measures 36 feet by 100 feet. At 90¢ per square yard, how much will it cost?

   Answer __10.80__

4. How many square feet are in a surface which measures 60 inches by 84 inches?

   Answer __15/20__

5. How many cubic feet are there in a box that measures 36 by 60 inches by 48 inches?

   Answer __103,680__

6. How many cubic yards of dirt have to be removed to make a basement measuring 25 ft by 27 ft by 16 ft?

   Answer __3,600__

7. How many cubic yards of dirt will be needed to fill a hole 27 ft by 20 ft by 10 ft?

   Answer __,800__

8. What is the capacity in cubic feet of a round tank whose diameter is 14 feet and whose height is 30 feet?

   Answer __4,618.2__

9. How many cubic feet of air are in a room that measures 12 feet by 9 feet by 8 feet?

   Answer __664__

10. How many cubic feet of water are needed to fill a pool 4 yards by 10 yards by 6 yards?

    Answer __240__

73

Change each measurement to the smaller unit.

1. $5\frac{1}{2}$ mi = _____ yd $3\frac{1}{2}$ ft = _____ in. $1\frac{1}{2}$ T = _____ lb

2. 17 m = _____ mm   5.7 km = _____ m   0.025 kg = _____ g

Change each measurement to the larger unit.

3. 90 in. = _____ ft   66 ft = _____ yd   29 oz = _____ lb

4. 3,500 mm = _____ m   19 m = _____ km   850 g = _____ kg

Find each answer.

5.
```
    1 0 lb  9 oz
  +   3 lb  8 oz
```
```
    1 1 gal  2 qt
  −   7 gal  3 qt
```
```
    6 yd  1 ft
  ×         9
```

6. $5\overline{)6\text{ T } 5\,0\,0\text{ lb}}$
```
    1 2 hr  2 5 min
  −   8 hr  4 0 min
```
```
    6 days  8 hr
  ×            6
```

Solve.

7. The path Jorge jogs forms a rectangle. The sides of the rectangle are 6 blocks and 9 blocks. How many blocks does Jorge jog?

   Answer _____

8. The sail on a boat is shaped like a triangle. It is 7 yards tall and is 4 yards along the base. How many yards of material are needed to make the sail?

   Answer _____

9. Karla made a puzzle that formed a triangle. The sides of the puzzle were 5 inches, 13 inches, and 12 inches. How many inches of wood does Karla need to make a frame for the puzzle?

   Answer _____

10. A rectangular-shaped toolbox is 24 inches long, 12 inches wide, and 8 inches tall. What is the volume of the toolbox?

    Answer _____

11. A circular-shaped garden has a radius of 14 feet. How much plastic edging will be needed to go around the garden?
    (Use $\pi = \frac{22}{7}$.)

    Answer _____

12. A window measures 1.5 meters by 2.3 meters. How much glass is needed for the window?

    Answer _____

Unit 3 Simple Equations, Ratio, Proportion, Powers, and Roots

Variables, Terms, Coefficients, and Monomials

To prepare for solving simple equations, it is important to learn the following vocabulary words.

Variables are letters or symbols used to represent numbers.

a	y	x^2	$-z$	c

A **term** is a number, or a variable, or a *product* or *quotient* of numbers and variables.

$\frac{a}{b}$	$-6x^2$	5	a	$8mn$

An **expression** consists of one or more terms separated by addition or subtraction.

$4 + b$	$ab - c$	$x + \frac{b}{t}$

Constants are terms that do not contain variables.

-7	2	$\frac{3}{4}$	0.8

The **coefficient** of a term is the numerical part. When there is no number in front of a variable, the coefficient 1 is understood. Coefficients may be fractions or decimals and may be positive ($+$) or negative ($-$).

Term:	$-7xy$	m	$-x$	$\frac{3}{4}x$ or $\frac{3x}{4}$
Coefficient:	-7	1	-1	$\frac{3}{4}$

A **monomial** is a number, a variable, or a *product* of a number and variables. So, the terms $\frac{s}{t}$ and $\frac{2}{3b}$ are *not* monomials because division is indicated by the variables in the denominator.

$9xy$	0.5	$-7k$	$\frac{2}{3}b$

Circle the constants.

1. -7 $4m$ 9 $\frac{1}{10}$ $-m$ 14 $x + y$ $2x$

Circle the monomials.

2. $5x$ $\frac{a}{b}$ $a + b$ $2xy$ $x - \frac{1}{2y}$ $\frac{1}{3}a$ $a - 2b$ $12z$

Circle the coefficient of each term.

3. $5a$ $4s$ $1.5f$ $-42b$ $1,250r$ $1.5n$ $-9m$ $\frac{2}{3}x$

Write the coefficient of each term.

4. $\frac{1}{2}x$ $\frac{-3}{4}y$ $\frac{z}{5}$ $\frac{3a}{2}$ $-\frac{2r}{5}$ $\frac{m}{4}$ $-a$ $\frac{3b}{5}$

_____ _____ _____ _____ _____ _____ _____ _____

75

Order of Operations

To **evaluate** an expression means to find a single value for it. If you were asked to evaluate $8 + 2 \cdot 3$, would your answer be 30 or 14? Since an expression has a unique value, a specific order of operations must be followed. The correct value of $8 + 2 \cdot 3$ is 14 because multiplication should be done before addition.

ORDER OF OPERATIONS
1. Do operations within parentheses.
2. Do all multiplications and divisions from left to right.
3. Do all additions and subtractions from left to right.

To change the expression so that it has a value of 30, write $(8 + 2) \cdot 3$. Now the order of operations is changed. Now the operation within the parentheses must be done first.

Sometimes you are given the value of a variable. You can evaluate the expression by substituting the value for each variable. Then perform the indicated operations.

$-8 \cdot 3$

EXAMPLE 1

Evaluate:
$12 - 2 \cdot 5 = 12 - 10 = 2$
$9(12 + 8) = 9(20) = 180$
$\frac{3 + 6}{1 + 2} = \frac{9}{3} = 3$
$\frac{1}{2}(3 + 11) = \frac{1}{2}(14) = 7$

EXAMPLE 2

Evaluate:
$r(r - 2)$ if $r = 6$.
$6(6 - 2) = 6(4)$
$\qquad = 24$

EXAMPLE 3

Evaluate:
$7p - \frac{1}{2}q$ if $p = 4$ and $q = 2$.
$7(4) - \frac{1}{2}(2) = 28 - 1$
$\qquad = 27$

Evaluate each expression.

1. $5 \cdot 2 + 1$ ⟨11⟩	$5(2 + 1)$ ⟨15⟩	$(4 + 6)(8)$ ⟨80⟩	$4 + 6 \cdot 8$ ⟨44 52⟩
2. $\frac{4 + 8}{3}$ ⟨4⟩	$\frac{7 + 9}{4}$ ⟨4⟩	$\frac{11(2) + 18}{8}$ ⟨7.25 5⟩	$\frac{10 - 2(3)}{2}$ ⟨2⟩
3. $\frac{1}{2}(6 + 26)$ ⟨16⟩	$\frac{1}{2}(6) + \frac{1}{2}(26)$ ⟨16⟩	$\frac{2}{3}(18) + 9$ ⟨21⟩	$\frac{2}{3}(18 + 9)$ ⟨18⟩

Evaluate each expression if $a = 9$, $b = 3$, and $c = 7$.

4. $4a + 7$ ⟨43⟩	$c(c + 3)$ ⟨70⟩	$7b + 2b$ ⟨27⟩	$(7 + 2)b$ ⟨27⟩
5. $2bc$ ⟨42⟩	$ab + ac$ ⟨90⟩	$a(b + c)$ ⟨90⟩	$ab + c$ ⟨34⟩
6. $\frac{a + b}{2}$ ⟨6⟩	$\frac{2a + 2b}{4}$ ⟨6⟩	$\frac{a}{b} + 9$ ⟨12⟩	$\frac{a + 9}{b}$ ⟨6⟩

Signed Numbers

A number line shows a set of positive and negative numbers called **integers.** All the numbers on the right side of zero are greater than (>) zero and are called **positive** numbers. All the numbers on the left side of zero are less than (<) zero and are called **negative** numbers.

$$\begin{array}{ccccccccccccccc} | & | & | & | & | & | & | & | & | & | & | & | & | & | & | \\ -7 & -6 & -5 & -4 & -3 & -2 & -1 & 0 & 1 & 2 & 3 & 4 & 5 & 6 & 7 \end{array}$$

Negative ← | → Positive

Use these rules when adding, subtracting, multiplying, or dividing integers.

Rule: To add numbers with the **same** sign, add and keep the same sign.
$5 + 4 = 9 \qquad -3 + (-1) = -4$
To add numbers with **different** signs, subtract and keep the sign of the larger.
$15 + (-4) = 11 \qquad -3 + 1 = -2$
Rule: To subtract numbers, change the sign of the number being subtracted and then use the rules for addition.
$-5 - (4) = -5 + (-4) = -9 \qquad -3 - (-1) = -3 + 1 = -2$
Rule: To multiply or divide numbers, first multiply or divide.
If numbers have the same sign, the answer is positive.

$5 \times 4 = 20 \qquad \frac{20}{4} = 5 \qquad -5 \times (-4) = 20 \qquad \frac{-20}{-4} = 5$

If numbers have different signs, the answer is negative.

$-5 \times 4 = -20 \qquad \frac{-20}{4} = -5 \qquad 5 \times (-4) = -20 \qquad \frac{20}{-4} = -5$

Solve.

1. $-14 + (+16) =$ +230	$25 + 64 =$ 89	$-31 + (-45) =$ -76
2. $6 + 8 + 4 =$ 18	$0 + (-5) + (-8) =$ -13	$-7 + 0 + (-8) =$ -15
3. $13 + (-17) =$ -30 ✗	$9 + (-12) =$ -21 ✗ -3	$27 + (-13) =$ +40 +14
4. $-3 + (-4) + 6 =$ 13 ✗	$-2 + 6 + 9 + (-4) =$ +15	$-10 + 7 + 3 =$ -20 ✗
5. $13 - 23 =$ 10	$-18 + 25 =$ -43	$-29 + 19 =$ 48 ✗
6. $18 - (-18) =$ +36	$25 - (-17) =$ +42	$-14 - 28 =$ +42 ✗
7. 43 $- (-29) =$ -14	$47 - 53 =$ +6 ✗	$-37 - (-45) =$ 8
8. $-4(-12) =$ +48	$-13(-7) =$ +91	$-5(23) =$ -115
9. $2(-1)(3) =$ -6	$3(-2)(-2) =$ +12 ✗	$-2(-4)(-3) =$ +24 ✗
10. $-126 \div 9 =$ -14	$234 \div (-13) =$ -18	$-240 \div (-15) =$ +16
11. $\frac{-18}{6} =$ -3	$\frac{24}{-8} =$ -3	$\frac{-40}{-5} =$ +8

Algebraic Expressions and Equations

To solve some types of word problems, you must know how to change word phrases into algebraic expressions. Then you can write a number sentence with a variable. This is called an equation.

EXAMPLE: Three times a number is thirty-six.

word phrase	algebraic term or expression	equation
Three times a number	$3n$	$3n = 36$
is	$=$	
thirty-six	36	

EXAMPLE: Maria has some money. Sue has five dollars less than Maria. Together they have forty-five dollars.

word phrase	algebraic term or expression	equation
Maria has some money.	m	$m + m - 5 = 45$
Sue has five dollars less than Maria.	$m - 5$	
Together they have forty-five dollars.	$= 45$	

Write algebraic terms or expressions for each part of the following problems. Then write the equation.

Do Your Work Here

1. Five more than a number is 25. _____

   $5 + n = 25$

   $5 + n = 25$

2. Twice a number divided by 7 is 14. _____

   _____

   $n^2 \div 7 = 14$ $2x \div 7 = 14$

3. Joe has fourteen more dollars than Jim. Together they have thirty-two dollars. _____

   _____

   $J + 14 = Jm$
   $J + Jm = 32$
   $x + x + 14 = 32$

4. Liz has seven times as many stamps as José has in his collection. Together they have three hundred twenty stamps. _____

   _____

   $5b = 35$ $b = 7$
   $\frac{5}{5}$ $\frac{35}{5}$

   $L \times 7 J = 120$

   $x + 7x = 320$

78

Missing Addends and Missing Factors

To **solve** an equation means to find the number that is a solution to the equation. The equations on this page involve missing addends or missing factors. Solve for a missing addend by **subtracting** the **same number** from **both sides** of the equation, as in Example 1. In Example 2, solve for a missing factor by **dividing**. The result in both examples shows the variable **isolated** on one side of the equal sign. (Notice that it does not matter which side.) The solution is on the other side. Check by substituting the solution into the original equation.

EXAMPLE 1

Solve: $r + 9 = 21$ r is the missing addend.

$r + 9 = 21$
$r + 9 - 9 = 21 - 9$ Check: $r + 9 = 21$
$r = 12$ $12 + 9 = 21$
The solution is 12. $21 = 21$

EXAMPLE 2

Solve: $99 = 9n$ n is the missing factor.

$99 = 9n$
$\frac{99}{9} = \frac{9}{9}n$
$11 = n$ Check: $99 = 9n$
   $99 = 9(11)$
The solution is 11. $99 = 99$

Solve. Check.

1. $b + 9 = 17$ $10 + a = 40$ $40 = 8y$ $12n = 48$

2. $2a = 8$ $5b = 35$ $n + 14 = 30$ $4x = 100$

3. $p + 33 = 333$ $7r = 77$ $b \cdot 6 = 90$ $17 + m = 83$

4. $180 = d \cdot 20$ $9 + n = 9$ $4 = 4y$ $f + 12 = 24$

5. $7x = 1$ $0 + k = 44$ $5z = 400$ $s + 50 = 555$

6. $6r = 3$ $7 + a = 8\frac{1}{2}$ $c + \frac{2}{3} = 5$ $2x = 6\frac{1}{2}$

Write an equation and solve.

Do Your Work Here

7. A number plus four is equal to ten. What is the number? _____

79

Combining Like Terms

8/6

Some equations contain the same variable in more than one term. Combine the like terms on each side of the equation. To combine like terms, add the coefficients of terms with the **same** variables. Then solve and check. Notice in the examples that sometimes you need to divide by a negative number.

EXAMPLE 1

Solve: $5a + 4a = 45$
$$9a = 45$$
$$\frac{9a}{9} = \frac{45}{9}$$
$$a = 5$$

Check: $5a + 4a = 45$
$$5(5) + 4(5) = 45$$
$$25 + 20 = 45$$
$$45 = 45$$

EXAMPLE 2

Solve: $3b - 5b = 4$
$$-2b = 4$$
$$\frac{-2b}{-2} = \frac{4}{-2}$$
$$b = -2$$

Check: $3b - 5b = 4$
$$3(-2) - 5(-2) = 4$$
$$-6 + 10 = 4$$
$$4 = 4$$

EXAMPLE 3

Solve: $4r + r = -5$
$$5r = -5$$
$$\frac{5r}{5} = \frac{-5}{5}$$
$$r = -1$$

Check: $4r + r = -5$
$$4(-1) + (-1) = -5$$
$$-4 + -1 = -5$$
$$-5 = -5$$

Solve. Check.

1. $-15c + 10c = 100$

$-5c = 100$
$\frac{-5c}{-5}$ $c = -20$

$4m + 7m = 66$

$m = 6$

$-2x + 3x = 24$

$x = 24$

$5z + 2z = 41 - 6$

$= 35$
$z = 5$

2. $8a - 5a = 60$

$a = 20$

$15a - 3a = 60$

$a = 5$

$-8x + 6x = 38$

$x = -9$

$-8z - 3z = 24 - 2$

$\frac{22}{5}$
$z = 2\frac{2}{5}$ $4\frac{2}{5}$

3. $14b - 6b = 48$

$b = 6$

$-6b + 5b = 55$

$b = 55$

$3x - x = 46$

$x = 23$

$3r + 2r = 31 - 6$

$r = 5$

Write an equation and solve.

4. Three times a number plus 4 times the same number is forty-two. What is the number?

$8z - 3z = 22$

$11z = 22$

$z = 2$

80

Solving Two-Step Equations

Some equations require two or more steps to solve.

Recall that when you evaluate an expression, the order of operations is multiplication and division before addition and subtraction. To isolate the variable in an equation, however, first add or subtract. Then multiply or divide.

EXAMPLE 1

Solve: $2x - 5 = 71$

Step 1

$2x - 5 + 5 = 71 + 5$
$2x = 71 + 5$
$2x = 76$

Step 2

$\frac{2x}{2} = \frac{76}{2}$
$x = 38$

Check: $2x - 5 = 71$
$2(38) - 5 = 71$
$71 = 71$

EXAMPLE 2

Solve: $4a + 7 = -49$

Step 1

$4a + 7 - 7 = -49 - 7$
$4a = -49 - 7$
$4a = -56$

Step 2

$\frac{4a}{4} = \frac{-56}{4}$
$a = -14$

Check: $4a + 7 = -49$
$4(-14) + 7 = -49$
$-49 = -49$

EXAMPLE 3

Solve: $-3y - 4 = 11$

Step 1

$-3y - 4 + 4 = 11 + 4$
$-3y = 11 + 4$
$-3y = 15$

Step 2

$\frac{-3y}{-3} = \frac{15}{-3}$
$y = -5$

Check: $-3y - 4 = 11$
$-3(-5) - 4 = 11$
$11 = 11$

Solve. Check.

1. $3x - 5 = 16$ $x = 7$

$-14r - 7 = 49$ $r = -4$

$25a - 4 = 96$ $A = 4$

$60y - 7 = 173$ $y = 3$

2. $7x + 3 = -4$ $x = -7$

$23y + 6 = 75$ X

$17a + 9 = 77$ X

$-2m + 9 = 7$ $m = 8$ $m = 1$

3. $8a - 7 = 65$ $A = 9$

$19b - 4 = 72$ $b = 4$

$52x + 4 = -100$ $x = -2$

$-6z + 15 = 3$ $z = -2$
$-6z + 15 = 3$
$-15 - 15$

Write an equation and solve.

4. Julio needs four times the amount he has now plus one hundred dollars to buy a car for $2,500. How much does he have now? $\underline{\quad 29 \quad}$ Ha

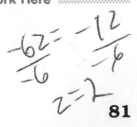

Do Your Work Here

$\frac{-6z}{-6} = \frac{-12}{-6}$
$z = 2$

81

Solving Equations with a Variable on Both Sides

Sometimes an equation will have variables and constants on both sides of the equal sign. You will need to rearrange the equation with all variables on one side and all constants on the other. To do this, use the same rules that you have been using as in the examples below.

EXAMPLE 1

Solve: $4x - 5 = 3x + 1$

$4x - 5 + (-3x) = 3x + 1 + (-3x)$ Add $-3x$ to both sides.
     $x - 5 = 1$ Combine terms.
    $x - 5 + 5 = 1 + 5$ Add $+5$ to both sides.
         $x = 6$ Combine terms.

Check: $4x - 5 = 3x + 1$
     $4(6) - 5 = 3(6) + 1$
        $19 = 19$

EXAMPLE 2

Solve: $7x - 1 = 15 + 3x$

$7x - 1 + 1 = 15 + 3x + 1$
     $7x = 16 + 3x$
  $7x - 3x = 16 + 3x - 3x$
     $4x = 16$
      $x = 4$

Check: $7x - 1 = 15 + 3x$
    $7(4) - 1 = 15 + 3(4)$
      $27 = 27$

Solve. Check.

1. $8x + 5 = 5x - 10$

$x = -5$

$5x - 1 = 4x + 3$

$x = 4$

$3x - 5 = x - 7$

$x = 12$

2. $5x - 16 = x + 4$

$x = 5$

$6x + 3 = 5x + 3$

$x = 0$

$15x + 5 = 10x - 15$

$x = -4$

3. $5x - 9 = 5 - 2x$

$\dfrac{-2x \quad +9 \quad +9 \quad -2x}{3x \qquad 14} \quad x = 12$

$5x + 20 = 4x - 24$

$x = -44$

$3x + 6 = x + 8$

$\dfrac{-6 \qquad\quad -6}{\qquad\qquad 4} \quad x = 2$

Write an equation and solve.

Do Your Work Here

4. Two times a number plus five is equal to three times the same number minus five. What is the number?

_____

79

Solving Equations with Parentheses

D/22

When an equation contains parentheses, remove the parentheses first. To remove parentheses, multiply the coefficients of each term within the parentheses by the number next to the parentheses. If the multiplier is negative, the sign of each term within the parentheses must be changed when the parentheses are removed. Remember, when there is no number next to the parentheses, the multiplier is 1.

EXAMPLE

Solve: $5b - 3(4 - b) = 2(b + 21)$

$5b - 12 + 3b = 2b + 42$

$8b - 12 = 2b + 42$

$8b - 12 - 2b = 2b + 42 - 2b$

$6b - 12 = 42$

$6b - 12 + 12 = 42 + 12$

$\frac{6b}{6} = \frac{54}{6}$

$b = 9$

Check: $5b - 3(4 - b) = 2(b + 21)$

$5(9) - 3(4 - 9) = 2(9 + 21)$

$60 = 60$

Solve. Check.

1. $3(x + 2) = 2(x + 5)$ $x = 4$	$3x - (2x - 7) = 15$ $x = 8$	$2b - 7(3 + b) = b + 3$ $x = -4$
2. $4(x - 3) = 2(x - 1)$ $x = 5$	$7x - (x - 1) = 25$ $x = 3$	$9a - 3(2a - 4) = 15$ $a = 1$
3. $4(x - 1) = 2(x + 4)$ $x = 2$ $x = 6$	$7 - 12(3 + b) = 31$ $b = -5$	$5x - 2(4 - x) = 20$ $x = 4$
4. $5x + 3 = 4(x + 2)$ $x = 11$	$5x - (x + 6) = 10$ $x = 1$	$4(2x - 5) = 3(x + 10)$ $x = 10$
5. $3(x - 1) = 2x + 3$ $x = 6$	$3x - (x + 2) = 4$ $x = 2$	$2(x + 5) - (x - 3) = 8$ $x = 1$

Pg 75

83

Using Equations

Write an equation and solve each problem.

1. Geraldo has three times as much money as does Pat. Together they have $12.80. How much does each have?

Let p = Pat's money.
Then 3p = Geraldo's money.
p + 3p = $12.80
4p = $12.80
p = $3.20 (Pat's money)
3p = $9.60 (Geraldo's money)

Geraldo ___$9.60___ Pat ___$3.20___

2. I am thinking of a number. Five times that number equals 240. What is the number?

Answer ___n=48___

3. Jane has five times as much money as does Helen. Together they have $42. How much does each have?

Jane ___35___ Helen ___7___

4. Ruth has $15 more than Julia. Together they have $55. How much does each have?

Ruth ___35___ Julia ___20___

5. One package contains 20 envelopes more than a second package. Together they contain 80 envelopes. How many envelopes does each package contain?

1st package ___20___ 2nd package ___60___

6. One lot contains 50 square feet less than a second lot. Together the lots contain 200 square feet. How many square feet are there in each lot?

1st lot ___125___ 2nd lot ___75___

7. Joe's mother is 3 times as old as Joe. The sum of their ages is 72 years. How old is each?

Mother ___54___ Joe ___18___

8. Frank's father is 2 years less than three times as old as Frank. The sum of their ages is 78 years. How old is each?

Father ___58___ Frank ___20___

84

Ordered Pairs

In the graph at the right, the horizontal number line labeled x is called the **x-axis.** The vertical number line labeled y is called the **y-axis.** The point where the axes cross is called the **origin** or zero.

The points on the graph are written as ordered pairs called **coordinates.** Ordered pairs are written in parentheses in the order (x, y). The first number is always the x-coordinate. The second number is always the y-coordinate.

The first number, x, tells how far the point is from the origin, O, on the x-axis. The second number, y, tells how far the point is from the origin, O, on the y-axis. In the graph, point A is -4 units to the left of the origin on the x-axis, and $+3$ units above the origin on the y-axis.

Point A is $(-4, 3)$.
Point B is $(4, 1)$.
Point C is $(-2, -3)$.
Point D is $(6, -6)$.

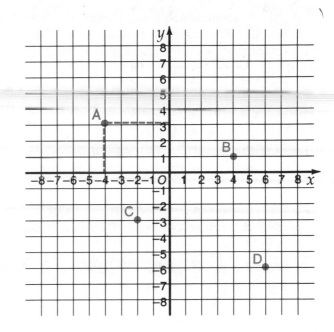

Find each point on the graph. Then write the ordered pairs.

1. P (_3_, _4_)   I (_0_, _2_)
2. E (_2_, _2_)   B (_-7_, _-6_)
3. D (_7_, _1_)   G (_0_, _-5_)
4. H (_-2_, _7_)   K (_-4_, _0_)
5. J (_-4_, _-2_) A (_3_, _7_)
6. C (_8_, _-6_)   F (_-7_, _2_)

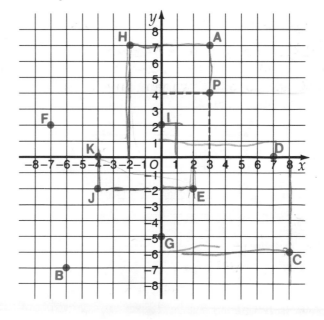

Graphing Ordered Pairs

To graph an ordered pair, locate the value of the x-coordinate on the x-axis. Then locate the value of the y-coordinate on the y-axis. Then follow the lines to where they meet and draw a dot or an x to plot the point. Then label the point.

To plot the point (2, −2), draw a dot below 2 on the x-axis, and across from −2 on the y-axis. This is point C.

To plot the point (5, 1), draw a dot above 5 on the x-axis, and across from 1 on the y-axis. This is point B.

To plot the point (−7, 2), draw a dot above −7 on the x-axis, and across from 2 on the y-axis. This is point A.

To plot the point (−4, −3), draw a dot below −4 on the x-axis, and across from −3 on the y-axis. This is point D.

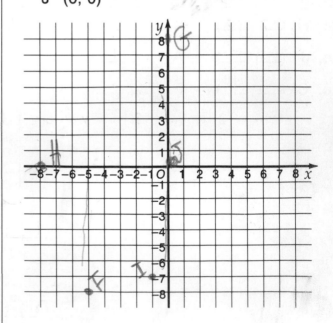

Plot and label each point on the graph provided.

1. A (2, 6)
B (3, −5)
C (−5, 3)
D (−4, −6)
E (0, 1)

2. F (−5, −8)
G (0, 8)
H (−8, 0)
I (0, −7)
J (0, 0)

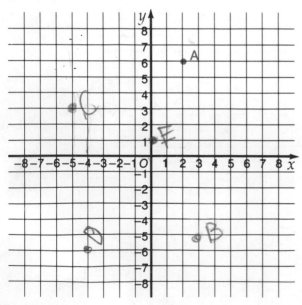

86

8/29

Simplify.

1. $(-2)(-7) =$ 14 $5(-8 + 9) =$ 5 $4 + 6(2) =$ 16

2. $-3a + 4(a + 2) =$ $7(-3a) + 6(-a) =$ 15 $3(a + 3b) - (a + 2b) =$
   $a + 8$ $-27a$

Solve. Check.

3. $5x = 90$ $x = 18$	$-2a = 4$ $A = -2$	$x - 10 = 35$ $x = 45$
4. $r + 17 = 20$ $r = 3$	$9y = -8$ $y = -1$ $-\frac{8}{9}$	$3x - 10 = 2$ $x = 4$
5. $8x + 5 = 5x - 10$ $x = 5$	$3y = 5$ $y = 2$ $y = 1\frac{2}{3}$	$3(x + 2) = 2(x + 5)$ $x = 4$
6. $3x + 4 = 5x - 12$ $x = 2$	$8a + 3 = 6a - 7$ $A = 2$ -5	$-2a = 2$ $A = -1$
7. $5x + 7 = 2x + 8$ $x = 3$	$3m + 1 = 6m - 5$ $m = 4$	$x + (x + 5) = x + 2$ $x = -3$
8. $a + 2(a - 2) = 2a + 3$ $a = 7$	$5y - 2(y + 2) = 5$ $y = 5$	$6y - 4 = 2(2y + 1)$ $y = -3$

Plot and label the ordered pairs.

9. A $(1, 1)$
   B $(-2, 3)$
   C $(-4, 2)$
   D $(3, -3)$
   E $(-2, -4)$

The Meaning of Ratio

A ratio is a fraction used to compare two quantities. For example, if a baseball player gets 3 hits for every 6 times at bat, the ratio of hits to times at bat is 3:6 or $\frac{3}{6} = \frac{1}{2}$. The ratio of times at bat to hits is 6:3 or $\frac{6}{3} = 2$.

Write each ratio.

1. The ratio of inches in a foot to inches in a yard

   Ratio _____ $12{:}36 \text{ or } \frac{12}{36} = \frac{1}{3}$ _____

   The ratio of hours in a day to hours in a week

   Ratio _____ $24{:}168 \quad \frac{24}{168}$ _____

2. The ratio of cups in a pint to cups in a quart

   Ratio _____ $2{:}4 \quad \frac{2}{4} = \frac{1}{2}$ _____

   The ratio of 3 apples on a table to 6 apples in a bowl

   Ratio _____ $3{:}6 \quad \frac{1}{2}$ _____

3. The ratio of cents in a quarter to cents in a dollar

   Ratio _____ $25{:}100 \quad \frac{25}{100}$ _____

   The ratio of 8 men to 10 women

   Ratio _____ $8{:}10 \quad \frac{8}{10}$ _____

4. The ratio of cents in a half dollar to cents in a dime

   Ratio _____ $10{:}50 \quad \frac{10}{50}$ _____

   The ratio of 10 women to 8 men

   Ratio _____ $10{:}8 \quad \frac{10}{8}$ _____

5. The ratio of 17 "yes" votes to 20 "no" votes

   Ratio _____ $17{:}20 \quad \frac{17}{20}$ _____

   The ratio of 5 wall outlets to 3 wall switches

   Ratio _____ $5{:}3 \quad \frac{5}{3}$ _____

6. The ratio of the number of days in December to the number of days in January

   Ratio _____ $31{:}31 \quad \frac{31}{31}$ _____

   The ratio of 8 hours asleep to 16 hours awake

   Ratio _____ $8{:}16 \quad \frac{8}{16}$ _____

7. The ratio of minutes in an hour to minutes in a half hour

   Ratio _____ $60{:}30 \quad \frac{60}{30}$ _____

   The ratio of 8 chicken legs to 8 chicken wings

   Ratio _____ $8{:}8 \quad \frac{8}{8}$ _____

8. The ratio of 10 bolts to 4 nuts

   Ratio _____ $10{:}4 \quad \frac{10}{4}$ _____

   The ratio of the number of tires on a bicycle to the number of tires on an automobile

   Ratio _____ $2{:}4 \quad \frac{2}{4}$ _____

Equations Involving Ratio

EXAMPLE: The perimeter of a rectangle is 300 in. The length and width are in the ratio of 3 to 2. Find the length and width.

Let $2x$ = width
$3x$ = length
Then, $2x + 2x + 3x + 3x = 300$ in.
$10x = 300$ in.
$x = 30$ in.
$2x = 60$ in. (the width)
$3x = 90$ in. (the length)

$2x$

$3x$

Solve these problems using ratio.

1. The length and width of a rectangle are in the ratio of 3 to 1. The perimeter is 80 feet. Find the length and width.

Length ___30___   Width ___10___

2. The perimeter of a triangle is 48 in. The sides are in the ratio of 1 to 2 to 3. Find the length of each side.

___8___   ___16___   ___24___

3. The perimeter of another triangle is 480 meters. The sides are in the ratio of 5 to 4 to 3. How long is each side?

___200___   ___160___   ___120___

4. Tom, Dick, and Harry together have $2.40, and the amount each one has is in the ratio of 1 to 2 to 3. How much does each have?

Tom ___40___   Dick ___80___   Harry ___120___

5. Maria, Jane, and Catherine together have $2.70. Their monies are in the ratio of 2 to 3 to 4. How much does each have? (Let $2x$ = Maria's money.)

Maria ___60___   Jane ___90___   Catherine ___120___

6. Three men contributed $84 to the Community Chest drive. Their contributions were in the ratio of 3 to 4 to 5. How much did each contribute?

1st man ___21___   2nd man ___28___   3rd man ___35___

7. Three persons gave $48 to the Red Cross. What the three persons gave was in the ratio of 3 to 4 to 5. How much did each give?

First ___9___   Second ___12___   Third ___15___

8. The population of a certain town is 3,600. The number of men, women, and children is in the ratio of 1 to 1.5 to 3.5. How many are in each of these three groups?

Men ___600___   Women ___600.5___   Children ___1800.5___

Probability

A probability (or chance) tells how likely it is that an event will occur. Probability in mathematics can be expressed as a ratio. The probability of an event occurring is the ratio of the number of favorable outcomes to the number of possible outcomes.

Probability of an event, $P(E) = \dfrac{\text{number of favorable outcomes}}{\text{number of possible outcomes}}$

EXAMPLE 1: Beth and Eduardo flip a coin to decide who will be the first to drive the new car. Eduardo flips the coin and Beth calls heads. What is the probability that Beth will win the toss?

There are two possible outcomes, heads or tails. There is one favorable outcome, heads.

Probability of heads $= \dfrac{1 \text{ (number of favorable outcomes)}}{2 \text{ (number of possible outcomes)}}$

The probability that Beth will win the toss is $\frac{1}{2}$.

EXAMPLE 2: There are 5 red marbles, 3 blue marbles, and 2 green marbles in a bag. If Maria picks one marble at random without looking, what is the probability that the marble will be blue?

$P(E) = \dfrac{3 \text{ the number of blue marbles}}{10 \text{ the total number of marbles}}$

The probability that the marble will be blue is $\frac{3}{10}$.

Write the probability of each event.

1. A bag contains 6 white marbles, 4 blue marbles, and 5 red marbles. What is the probability that a red marble will be picked at random?

   $\frac{5}{15} = \frac{1}{3}$

   Answer _____

2. A number cube has six sides. Each side has a number on it from 1 to 6. If a cube is rolled once, what is the probability that a 5 will show?

   $\frac{1}{6}$

   Answer _____

3. Wong has 3 shirts in his drawer, 1 blue, 1 white, and 1 black. If he takes a shirt without looking, what is the probability that it will be black?

   $\frac{1}{3}$

   Answer _____

4. Betty has a $1 bill, a $5 bill, a $10 bill, and a $20 bill in her purse. If she randomly draws out one of the bills, what is the probability that it will be a $10 bill?

   $\frac{1}{4}$

   Answer _____

5. What is the probability of randomly drawing a can of peas from a box that contains 3 cans of beans, 8 cans of peas, and 5 cans of corn?

   $\frac{8}{16} = \frac{1}{2}$

   Answer _____

6. Each card in a stack has a letter on it. The letters spell the word *carrot*. What is the probability of randomly drawing a card with the letter *r*?

   $\frac{2}{6} = \frac{1}{3}$

   Answer _____

Solve.

1. Mary has $2.50. Jane has $2. What is the ratio of Mary's money to Jane's?

   Answer _____ 250:2 _____

2. What is the ratio of Jane's money to Mary's?

   Answer _____ 2:2,50 _____

3. Trenton has a population of 40,000 persons. There are 6,000 children of school age in Trenton. What is the ratio of the school children to the total population?

   Answer _____ 40000 : 6000 _____ $\frac{6}{40}$ $\frac{3}{20}$

4. If a coin is flipped once, what is the probability that the result will be tails?

   Answer _____ $\frac{1}{2}$ _____

5. If a bag contains 2 red, 3 blue, and 4 white marbles, what is the probability that a random pick will be a white marble?

   Answer _____ $\frac{4}{9}$ _____

6. Tai, Bill, and James together have $8.40. Bill has twice as much as Tai. James has twice as much as Bill. How much does each have?

   Tai _____ Bill _____ James _____

7. If a coin is flipped once, what is the probability that the result will be heads?

   Answer _____ $\frac{1}{2}$ _____

8. John has a $1 bill, a $5 bill, and a $10 bill in his pocket. What is the probability that he will randomly choose a $5 bill?

   Answer _____ $\frac{1}{3}$ _____

9. A bag contains 5 cans of beans, 6 cans of corn, and 8 cans of peas. What is the probability that Sally will randomly pick a can of corn?

   Answer _____ $\frac{6}{19}$ _____

10. A number cube has 6 sides. Each side has a number from 1 to 6. What is the probability of rolling a 4 on one roll of the cube?

    Answer _____ $\frac{1}{6}$ _____

11. There are 125 students and 5 teachers in the 10th grade at Sabrina's school. What is the ratio of teachers to students?

    Answer _____ 5:125 _____ $\frac{5}{125}$ $\frac{1}{25}$

12. Each card in a stack has a letter on it. When the letters are put together, they spell the word *cannon*. What is the probability that a card drawn at random will have an *n* on it?

    Answer _____ $\frac{3}{6}$ _____ $\frac{1}{2}$

13. The sum of three numbers is 180. The three numbers are in the ratio of 1:2:3. What are the three numbers?

    _____ _____ _____

14. The ratio of peanuts to pecans to cashews in a mixture is 5:3:1. Per pound, peanuts cost $0.50, pecans $0.86, and cashews $1.20. How much would 9 lb of this mixture cost?

    Answer _____

Using Ratio and Probability

1. Fred Aarons is 6 ft tall, and his wife is 64 in. tall. What is the ratio of Fred's height to that of his wife? (Remember to change to the same units—use inches.)

   Answer __72 : 64__

2. The ratio of Jim's age to his father's age is 2 : 5. If the total of their ages is 70, how old is Jim's father?

   Answer _____

3. In problem 1, what is the ratio of the wife's height to Fred's?

   Answer __64 : 72__

4. The diameter of a certain circle is 16 inches. The diameter of a smaller one is 4 inches. What is the ratio of the radius of the larger circle to that of the smaller circle?

   Answer _____

5. The total paid attendance at a recent high school football game was 6,000. If the ratio of the number of adults to the number of students was 7 : 5, how many adults attended the game?

   Answer _____

6. The ratio of the height of a certain elm tree to that of a nearby oak tree is 3 : 2. If the total of the tree heights is 100 ft, how tall is the oak?

   Answer _____

Write the probability for each of the following.

7. A circle is divided into 8 equal parts. Each part is numbered from 1 to 8. What is the probability that a spinner will land on the number 7 in one try?

   Answer __$\frac{1}{8}$__

8. What is the probability that the spinner in Exercise 7 will land on an even number in one try?

   Answer _____

9. There are 10 cans in a box. Half the cans contain red paint and half contain green paint. What is the probability that Yolanda will randomly draw a can of green paint?

   Answer __$\frac{5}{10}$__

10. Alex has 3 black shirts, 5 blue shirts, and 2 brown shirts in a drawer. What is the probability that he will randomly pick a brown shirt?

    Answer _____

11. Agnes has 20 marbles in a bag. There are 5 blue marbles, 10 green marbles, 3 red marbles, and 2 orange marbles. If Agnes randomly draws one marble, what is the probability that it will be red?

    Answer __$\frac{3}{20}$__

12. What is the probability that a marble randomly picked from the bag in Exercise 11 will be green?

    Answer _____

The Meaning of Proportion

When two ratios are **equal,** they are said to form a **proportion.** Thus, the ratios 1 to 2 and 8 to 16 form a proportion. Proportion is usually written in this form: $1:2 = 8:16$, and it is read, "1 is to 2 **as** 8 is to 16."

The first and last terms of a proportion are called the **extremes,** and the second and third terms are called the **means.** The product of the means equals the product of the extremes. This is true in all proportions.

extreme:mean = mean:extreme

Study this proportion. 1 and 16 are the extremes, 2 and 8 are the means.

EXAMPLE: 1:2 = 8:16

The product of the means equals the product of the extremes.
$$1 \times 16 = 2 \times 8$$
$$16 = 16$$

This rule may be used to solve a proportion in which three of the members are known and the fourth is unknown.

EXAMPLE 1	EXAMPLE 2	EXAMPLE 3	EXAMPLE 4
extremes means $5:10 = 10:x$ $5x = 100$ $x = 20$	$25:30 = 10:x$ $25x = 300$ $x = 12$	$5:12 = 10:x$ $5x = 120$ $x = 24$	$7:9 = 21:x$ $7x = 189$ $x = 27$

Solve these proportion problems.

1. $4:15 = 8:x$
  $4x = 120$
  $x = 30$

$5:12 = 10:x$
  $x = 24$

$5:x = 15:9$
  $x = 3$

$20:x = 5:4$
  $x = 16$

2. $100:x = 25:3$
  $x = 12$

$12:4 = x:7$
  $x = 21$

$6:9 = x:3$
  $x = 2$

$7:3 = x:6$
  $x = 14$

3. $x:4 = 5:10$
  $x = 2$

$x:20 = 4:5$
  $x = 16$

$x:100 = 3:25$
  $x = 12$

$x:6 = 2:3$
  $x = 4$

4. $x:3.2 = 5:8$
  $x = 2$

$6:7 = x:14$
  $x = 12$

$x:3 = 10:15$
  $x = 2$

$7.5:x = 2:4$
  $x = 150$

Proportions as Fractions

Proportions are often written in the form of fractions. The proportion $4:5 = 8:x$ can be written $\frac{4}{5} = \frac{8}{x}$. Using the rule, **the product of the means equals the product of the extremes:** $4x = 40$, or $x = 10$. The numerator of one fraction is multiplied by the denominator of the other fraction. This procedure is called **cross multiplication**.

(handwritten at top right:) $\frac{9}{25}$ $\frac{x}{3} = \frac{10}{5}$ $x = 6$ $5x = 30$

EXAMPLE 1

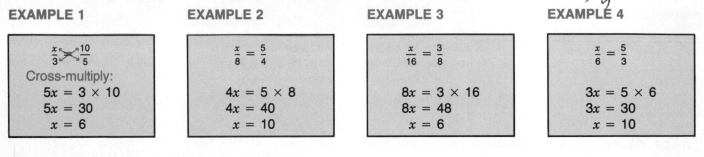

$$\frac{x}{3} \diagdown \frac{10}{5}$$

Cross-multiply:
$5x = 3 \times 10$
$5x = 30$
$x = 6$

EXAMPLE 2

$$\frac{x}{8} = \frac{5}{4}$$

$4x = 5 \times 8$
$4x = 40$
$x = 10$

EXAMPLE 3

$$\frac{x}{16} = \frac{3}{8}$$

$8x = 3 \times 16$
$8x = 48$
$x = 6$

EXAMPLE 4

$$\frac{x}{6} = \frac{5}{3}$$

$3x = 5 \times 6$
$3x = 30$
$x = 10$

Solve each proportion. Use cross multiplication.

1. $\frac{x}{9} = \frac{1}{3}$

   $3x = 9$
   $x = 3$

   $\frac{x}{3} = \frac{2}{5}$

   $\frac{5x = 6}{5} \Big/ \frac{}{5}$ $x = \frac{6}{5} = 1\frac{1}{5}$

   $\frac{x}{12} = \frac{1}{6}$

   $x6 = 12$
   $x = 2$

   $\frac{x}{4} = \frac{3}{2}$

   $2x = 12$
   $x = 6$

2. $\frac{2x}{3} = \frac{10}{5}$

   $2x \cdot 5 = 30$
   $x = 3$

   $\frac{x}{5} = \frac{7}{5}$

   $5x = 35$
   $x = 7$

   $\frac{8}{x} = \frac{4}{5}$

   $x4 = 40$
   $x = 10$

   $\frac{5}{x} = \frac{7}{14}$

   $x7 = 70$
   $x = 10$

3. $\frac{6}{x} = \frac{3}{5}$

   $3x = 30$
   $x = 10$

   $\frac{4}{x} = \frac{2}{3}$

   $2x = 12$
   $x = 6$

   $\frac{3}{x} = \frac{5}{2}$ $5x = 6$

   $5x = 6$
   $x = 1\frac{1}{5}$

   $\frac{16}{x} = \frac{8}{3}$

   $8x = 48$
   $x = 6$

4. $\frac{15}{9} = \frac{5}{x}$

   $15x = 45$
   $x = 3$

   $\frac{5}{3} = \frac{25}{x}$

   $5x = 75$
   $x = 15$

   $\frac{7}{6} = \frac{14}{x}$

   $7x = 84$
   $x = 12$

   $\frac{5}{5} = \frac{13}{x}$

   $5x = 65$
   $x = 13$

Using Proportion

It is important to remember that the order of comparison must be the same in both the first and the second ratios. If the first ratio compares dollars to time, then the second ratio must also compare dollars to time in the same order.

EXAMPLE: If Bob earns $9 in 6 days on his paper route, at the same rate how much will he earn in 2 days?

The proportion from which this problem may be solved can be written in a number of different ways:

$$\frac{\text{dollars}}{\text{time}} = \frac{\text{dollars}}{\text{time}} \text{ or } \frac{9}{6} = \frac{x}{2} \qquad \frac{\text{time}}{\text{dollars}} = \frac{\text{time}}{\text{dollars}} \text{ or } \frac{6}{9} = \frac{2}{x}$$

$$\frac{\text{dollars}}{\text{dollars}} = \frac{\text{time}}{\text{time}} \text{ or } \frac{9}{x} = \frac{6}{2} \qquad \frac{\text{dollars}}{\text{dollars}} = \frac{\text{time}}{\text{time}} \text{ or } \frac{x}{9} = \frac{2}{6}$$

To solve, use any of these proportions.

Cross multiply: $6x = 2 \times 9$ Check: $\frac{6}{9} = \frac{2}{3}$
$$6x = 18$$
$$x = 3 \qquad\qquad 18 = 18$$

Bob will earn $3 in 2 days.

Solve these problems involving proportion.

1. If Paul can walk 15 miles in 6 hours, how far can he walk, at the same rate, in 8 hours?

   Answer $x = 20$ $\dfrac{15}{6}\ \dfrac{x}{8}$

2. In the 1972 Olympic Games the Soviet Union's Valeri Borzov won first place in the 100-meter dash with a time of 10.1 seconds. At this same rate, how long would it take him to run 800 meters?

   Answer $x = 80.8$ $\dfrac{100}{10.1}\ \dfrac{800}{x}$

3. Ann runs the 50-yard race in 6 seconds. At the same speed, how long will it take her to run 75 yards?

   Answer $x = 9$ $\dfrac{50}{6}\ \dfrac{75}{x}$

4. Marie saved $50 in 4 weeks. At the same rate how much will she save in 6 weeks?

   Answer $x = 75$ $\dfrac{50}{4}\ \dfrac{x}{6}$

5. Joe spent $4.36 for gasoline in driving 68 miles. How much would he spend in driving 85 miles?

   Answer $x = 5.45$ $\dfrac{68}{4.36}\ \dfrac{85}{x}$

6. The taxes on a piece of property valued at $4,000 were $66. At this same rate, what would be the taxes on property valued at $5,500?

   Answer $x = 90.75$ $\dfrac{4,000}{66}\ \dfrac{5,500}{x}$

7. A car is traveling at the rate of 45 miles per hour. At the same speed, how far will it travel in 40 minutes? (Change the 1 hour to minutes.)

   Answer $x = 30$ $\dfrac{45}{60}\ \dfrac{x}{40}$

8. Frank Budd's best time in running 100 yards is 9.2 seconds. At this same rate, how many seconds would it take him to run 220 yards?

   Answer $x = 20.24$ $\dfrac{100}{9.2}\ \dfrac{220}{x}$

95

Proportion in Area and Volume

Remember that a square is a rectangle with all sides equal.

EXAMPLE: Notice the figures at the right. If square *P* has a side of 1 in., and square *Q* has a side of 2 in., the area of *Q* is how many times the area of *P*?

Write a proportion using the area formula $A = s \times s$, or s^2.

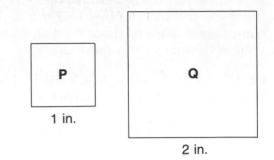

P

Q

1 in.

2 in.

> Using proportion, we have:
> $Q:P = 2^2:1^2$
> $Q.P - 4:1$
> The area of Q is 4 times that of P.

Solve each problem below.

---

1. A cube is a prism having equal length, width, and height. The volume of a cube is $s \times s \times s$, or s^3 (read, "s cubed").
One cube has a side of 1 in., and a second cube has a side of 2 in. The volume of the second cube is how many times that of the first cube?

$$V_1:V_2 = 1^3:2^3$$
$$V_1:V_2 = 1:8$$

Answer _____ 8 times _____

2. A mixture of 2 parts sand and 3 parts gravel is used to make 25 cubic yards of concrete. How many cubic yards of each are needed? (Hint: Solve two proportions.)

*2 parts + 3 parts = 5 parts.
Then 2:5 = x:25 (sand)
and 3:5 = x:25 (gravel).*

sand / gravel

Answer __ 10:25 / 15:25 __

---

3. One square has a side of 3 in., and a second square has a side of 6 in. The area of the second is how many times that of the first?

Answer _____ 2 _____

4. One square has a side of $2\frac{1}{2}$ times the side of a second square. The area of the first is how many times that of the second? (Change $2\frac{1}{2}$ to 2.5.)

Answer _____ 6.25 _____

---

5. The side of one cube is 2 in., and the side of a second cube is 3 in. The volume of the second cube is how many times that of the first? (Do not be afraid of fractions in the answer.)

Answer _____ 3.3 _____

6. A picture 7 in. by 12 in. is enlarged so that the 7 in. side will be 14 in. What will be the other dimension of the enlarged picture?

Answer __ 14 by 24 __

---

7. The side of one cube is 2 in., and the side of a second cube is 4 in. The volume of the second cube is how many times that of the first?

Answer _____ ~~4~~ 8 _____

8. In making solder, 2 parts of tin and 3 parts of lead are used. How much of each is used in making 65 pounds of solder?

Answer __ 24 31 __

96

Similar Triangles

Two triangles are *similar* if the corresponding angles are the same size and if the ratio of the lengths of corresponding sides are equal. Since the ratios are equal, they form a proportion. Similar triangles are the same shape.

If you are given the lengths of some of the sides of two similar triangles, you can find the lengths of other sides. To do this, set up and solve proportions.

EXAMPLE: In similar triangles ABC and XYZ, AB = 5 m, XY = 6 m, and BC = 10 m. What is the length of YZ?

Write a proportion.	$\frac{AB}{XY} = \frac{BC}{YZ}$	or	$\frac{AB}{BC} = \frac{XY}{YZ}$
	$\frac{5}{6} = \frac{10}{YZ}$		$\frac{5}{10} = \frac{6}{YZ}$
Cross multiply and solve.	$5YZ = 60$		$5YZ = 60$
	$YZ = 12$ m		$YZ = 12$ m

Set up and solve a proportion to solve each problem.

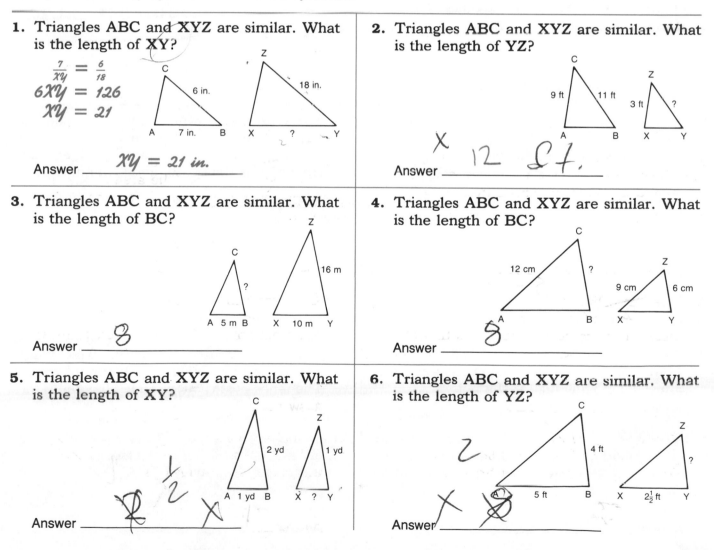

1. Triangles ABC and XYZ are similar. What is the length of XY?

$\frac{7}{XY} = \frac{6}{18}$

$6XY = 126$

$XY = 21$

Answer ___$XY = 21$ in.___

2. Triangles ABC and XYZ are similar. What is the length of YZ?

Answer ___12 ft.___

3. Triangles ABC and XYZ are similar. What is the length of BC?

Answer ___8___

4. Triangles ABC and XYZ are similar. What is the length of BC?

Answer ___8___

5. Triangles ABC and XYZ are similar. What is the length of XY?

Answer _____

6. Triangles ABC and XYZ are similar. What is the length of YZ?

Answer _____

97

Using Proportion in Similar Figures

You can use proportions to solve problems about similar figures.

To find the height of the tree shown here, use similar triangles to form a proportion. Then solve.

10/16

EXAMPLE: A tree casts a shadow 60 feet long. At the same time, an 8-foot post casts a 12-foot shadow. How tall is the tree?

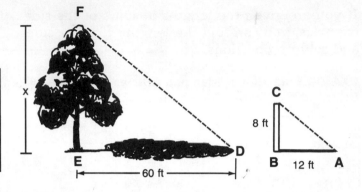

$\frac{x}{60} = \frac{8}{12}$	Write a proportion.
$12x = 8(60)$	Cross multiply.
$\frac{12}{12}x = \frac{8(60)}{12}$	
$x = 40$	The tree is 40 feet tall.

Complete the proportions, referring to triangles ABC and DEF above.

1. $\frac{EF}{DE} = \frac{BC}{\quad}$ $\frac{DF}{\quad} = \frac{DE}{AB}$

2. $\frac{DE}{AB} = \frac{\quad}{AC}$ $\frac{AC}{\quad} = \frac{AB}{DE}$

Draw a picture. Use a proportion. Solve.

3. A flagpole casts a 75-foot shadow at the same time a tree 20 feet tall casts a shadow of 30 feet. How tall is the flagpole?

    $4 \times 9 \quad \frac{12}{36}$

    50

    20

    Answer _____

4. The Washington Monument casts a shadow 111 feet long at the same time a 50-foot telephone pole casts a shadow 10 feet long. How high is the Washington Monument?

    555

    $22 \quad 4R56$

    Answer _____

5. A telephone pole casts a shadow 30 feet long while a fence post 4 feet high casts a shadow 3 feet long. How high is the pole?

    40

    Answer _____

6. A smokestack casts a shadow of 40 feet while a fence post nearby casts a shadow of 2 feet. The fence post is 5 feet high. How tall is the smokestack?

    100 ✓

    Answer _____

Checking Up

Solve each of the following proportions to find the value of the letter shown.

1. $\frac{9}{x} = \frac{3}{4}$ $x=12$

$\frac{6}{5} = \frac{x}{10}$ $x=12$

$\frac{x}{12} = \frac{3}{2}$ $x=18$

$\frac{17}{10} = \frac{8.5}{x}$ $x=5$

2. $\frac{6}{11} = \frac{3}{x}$ $x=5.5$

$\frac{9}{2} = \frac{36}{x}$ $x=8$

$\frac{15}{40} = \frac{x}{8}$ $x=3$

$\frac{35}{14} = \frac{50}{x}$ $x=20$

3. $\frac{20}{19} = \frac{30}{x}$ 28.5

$\frac{27}{x} = \frac{15}{10}$ $x=18$

$\frac{3}{9} = \frac{x}{15}$ $x=5$

$\frac{21}{75} = \frac{x}{25}$ $x7$

4. $5:1 = 30:x$ $x=6$

$x:21 = 2:3$ $x=14$

$12:x = 16:4$ $x=3$

$7:x = 28:8$ $x=2$

5. $12:6 = 8:x$ $x=4$

$8:x = 32:12$ $x=3$

$4:x = 16:12$ $x=3$

$35:14 = 50:x$ $x=20$

Solve.

6. Triangles **ABC** and **XYZ** are similar. What is the length of BC?

7. Triangles **ABC** and **XYZ** are similar. What is the length of YZ?

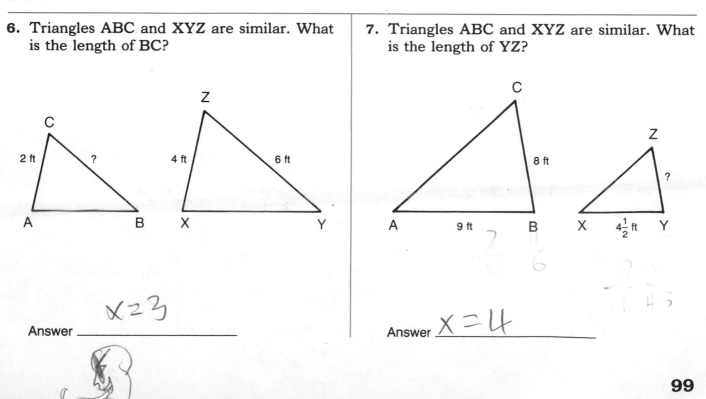

$x=3$

Answer _____

$x=4$

Answer _____

99

Exponents

An exponent is a number placed above and to the right of another number. For example, in the expression y^2, the 2 is the exponent. The y is called the base.

An exponent tells how many times a number is used as a factor. Therefore, $y \times y = y^2$.

Remember, $y \times y$, $y \cdot y$, and $(y)(y)$ all mean multiply.

EXAMPLE 1

Write $(7)(7)(7)(7)(7)$ using an exponent.
$(7)(7)(7)(7)(7) = 7^5$

EXAMPLE 2

Write $r \cdot r \cdot r \cdot r$ using an exponent.
$r \cdot r \cdot r \cdot r = r^4$

EXAMPLE 3

Write y using an exponent.
$y = y^1$

Write using exponents.

1. $5 \times 5 = 5^2$	$4 \times 4 \times 4 =$ 4^3	$(y)(y) =$ y^2	$x \cdot x =$ x^2
2. $2 \times 2 \times 2 \times 2 =$ 2^4	$a \cdot a \cdot a =$ a^3	$g \cdot g \cdot g \cdot g =$ g^4	$(6)(6)(6)(6) =$ 6^4
3. $9 \times 9 \times 9 =$ 9^3	$(r)(r)(r) =$ r^5	$p =$ p	$2 \times 2 =$ 2^2
4. $s \cdot s \cdot s \cdot s \cdot s =$ 8^5	$1 \times 1 \times 1 \times 1 =$ 1^4	$15 \times 15 =$ 15^2	$t \cdot t \cdot t =$
5. $(3)(3)(3) =$	$w \cdot w =$	$(5)(5)(5) =$	$7 \times 7 \times 7 \times 7 =$
6. $(a)(a) =$	$(9)(9)(9)(9) =$	$z \cdot z \cdot z \cdot z =$	$q =$
7. $4 \times 4 =$	$3 \cdot 3 \cdot 3 =$	$x =$	$(4)(4)(4)(4) =$

Powers of Numbers

The exponent in an expression indicates the number of times the base is used as a factor. For example, $2^3 = 2 \cdot 2 \cdot 2 = 8$.

When a base with an exponent also has an exponent, such as $(2^2)^2$, simplify by multiplying the two exponents. $(2^2)^2 = 2^{2 \cdot 2} = 2^4$. You can check multiplication by showing the factors and adding the exponents. $(2^2)^2 = 2^2 \cdot 2^2 = 2^{2+2} = 2^4 = 16$

Remember,
- zero with any exponent is zero. $0^2 = 0 \cdot 0 = 0$
- one with any exponent is one. $1^2 = 1 \cdot 1 = 1$

EXAMPLE 1

Simplify: 4^3
$4^3 = 4 \cdot 4 \cdot 4$
$= 64$

EXAMPLE 2

Simplify: $(2)^4$
$(2)^4 = (2)(2)(2)(2)$
$= 4 \cdot 4$
$= 16$

EXAMPLE 3

Simplify: $2^4 \cdot 3^2$
$2^4 \cdot 3^2 = (2 \cdot 2 \cdot 2 \cdot 2)(3 \cdot 3)$
$= 16 \cdot 9$
$= 144$

EXAMPLE 4

Simplify: $(2^2)^3$
$(2^2)^3 = 2^{2 \cdot 3} = 2^6 = 64$

Simplify.

1. $4^2 = 16$ | $3^2 =$ $\quad 3 \times 3 = 9$ | $2^2 =$ $\quad 2 \cdot 2 = 4$ | $5^2 =$

2. $8^2 =$ | $0^3 =$ | $(10)^2 =$ | $9^2 =$

3. $(1)^3 =$ | $10^3 =$ | $2^5 =$ | $(4)^4 =$

4. $(11)^3 =$ | $10^1 =$ | $21^2 =$ | $3^6 =$

5. $14^2 =$ | $20^3 =$ | $17^2 =$ | $0^{50} =$

6. $2^3 \cdot 3^2 =$ | $4^2 \cdot 5^2 =$ | $(3)^3 \cdot 4^2 =$ | $8^2 \cdot 2^3 =$

7. $(10^2)(5^3) =$ | $(9)^2(3^3) =$ | $(6^2)(7^3) =$ | $(1)^2(5^2) =$

8. $(3^2)^3 =$ | $(4^2)^2 =$ | $(5^2)^3 =$ | $(6^3)^2 =$

9. $(3^4)(2^2)^2 =$ | $(0^3)^5(4^3)^5 =$ | $(3^2)^2(1)^4 =$ | $(2^3)^2(2^2)^2 =$

10/18

Square Roots

Finding square roots is the opposite of finding powers of numbers. The sign $\sqrt{\ }$ tells you to find the positive square root of the number under the sign. The expression $\sqrt{9}$ is called a **radical**. The 9 is called a **radicand** and the $\sqrt{\ }$ is called a **radical sign**.

You may remember from arithmetic that the square root of a number is that factor which when multiplied by itself gives the number. Here are some examples:

| $\sqrt{4} = 2$ because $2 \cdot 2 = 4$ | $\sqrt{9} = 3$ because $3 \cdot 3 = 9$ |

You may also take the square root of a monomial. To do this, first take the square root of the coefficient. Next, take the square root of each variable.

| $\sqrt{4x^2} = 2x$ | $\sqrt{9x^4} = 3x^2$ | $\sqrt{x^4y^2} = x^2y$ |
| Check: $2x \cdot 2x = 4x^2$ | Check: $3x^2 \cdot 3x^2 = 9x^4$ | Check: $x^2y \cdot x^2y = x^4y^2$ |

Use the chart to help you find larger square roots. To find $\sqrt{400}$, find 400 in the *square* column. The square root of 400 is in the first column. So, $\sqrt{400}$ is 20 since $20 \times 20 = 400$.

Remember, the square root of 1 is always 1.

Number	Square
1	1
2	4
3	9
4	16
5	25
6	36
7	49
8	64
9	81
10	100
11	121
12	144
13	169
14	196
15	225
16	256
17	289
18	324
19	361
20	400

Find each square root.

1. $\sqrt{121} = $ *11* $\sqrt{49} = $ 7 $\sqrt{144} = $ 12 $\sqrt{25} = $ 5

2. $\sqrt{225} = $ 15 $\sqrt{81} = $ 9 $\sqrt{16} = $ 4 $\sqrt{324} = $ 18

3. $\sqrt{a^2} = $ a $\sqrt{a^4} = $ A^2 $\sqrt{x^6} = $ x^3 $\sqrt{y^8} = $ y^4

4. $\sqrt{a^2b^2} = $ $A \cdot b$ $\sqrt{x^4y^4} = $ $x^2 \cdot y^2$ $\sqrt{x^2y^4} = $ xy^2 $\sqrt{a^2b^2c^2} = $ abc

5. $\sqrt{9a^2} = $ $3a$ $\sqrt{16b^4} = $ $4b^2$ $\sqrt{36a^2b^2} = $ $6ab$ $\sqrt{64x^2y^4} = $ $8xy^2$

6. $\sqrt{169x^{10}y^4} = $ $13x^5y^2$ $\sqrt{289x^{12}} = $ $17x^6$ $\sqrt{324b^4c^6} = $ $18b^2c^3$ $\sqrt{400a^2b^6} = $ $20ab^3$

102

Problem-Solving Strategy: Identify Extra Information

Some problems contain information that is not needed to solve the problem. When you read a problem, it may be helpful to cross out the extra facts.

STEPS

1. Read the problem.

Everett typed a 500-word essay in 6 minutes 30 seconds. How many seconds did it take him to type the essay?

2. Decide which facts are needed.

The problem asks for the number of seconds. You need to change 6 minutes to seconds, then add 30 seconds. Recall that there are 60 seconds in a minute.

3. Decide which facts are extra.

You do not need to know that the essay has 500 words.

4. Solve the problem.

Multiply the number of minutes by 60. Then add 30 seconds.

Seconds in 6 minutes: 6 × 60 = 360 seconds

Total seconds: 360 + 30 = 390 seconds

Everett typed the essay in 390 seconds.

In each problem, cross out the extra facts. Then solve the problem.

1. Oliver worked 8.5 hours on Monday, 6 hours on Tuesday, ~~and 7.5 hours on Wednesday.~~ How much longer did he work on Monday than Tuesday?

Answer _____

2. Winona bought a car that cost $5,480, which includes a $199 service warranty. She paid $1,000 as a down payment and took out a loan for the rest. What was the amount of her loan?

Answer _____

3. Ernest works part-time and earns $75 per week. If he saves about $25 per week, how much money will he save in 6 weeks?

Answer _____

4. Iris spent $4.32 for a calculator, $12.89 for a backpack, and $8.99 for a lantern. How much more did the lantern cost than the calculator?

Answer _____

103

Applying Your Skills

Solve.   10/23

1. Complete this proportion:
   $7:15 = 21:x$

   Answer $X=45$

2. Find the missing quantity in this proportion:
   $x:10 = 15:30$

   Answer $x=5$

3. In the similar triangles ABC and DEF, AB is 2 in., BC is 3 in., and DE is 4 in. Find the length of EF. (Hint: Make a diagram.)

   Answer $EF = 6$

4. In the same similar triangles find AC if DF is 5 in.

   $Df = 10$ 2.5

   Answer

5. The side of a square is 1 in. The side of a second square is 2 in. The area of the second square is how many times that of the first?

   Answer 4

6. Find the value of x in these proportions:

   $\frac{10}{20} = \frac{x}{16}$

   $\frac{x}{15} = \frac{10}{3}$

   Answer $x=8$ $x=50$

7. Rosa earned $46 in 12 hours. By using proportion, show how much she would earn in 15 hours.

   Answer $57.50

8. The Empire State Building casts a shadow 250 feet long, while a nearby 100-foot flagpole casts a shadow 20 feet long. How tall is the Empire State Building?

   $x = 50$

   Answer

   $\frac{100}{20}$ $\frac{250}{x}$

Solve.

9. Triangle ABC is similar to $\triangle$XYZ. What is the length of XY?

   $? = 12$

   Answer

104

9/1

Solve. Simplify.

1. $6 = 2x - 20$ $3x = 63$ $X=21$ $8x + 5 = 29$ $6x - 9 = 21$

$X=-13$ $x=3$ $X=20$ $X=5$

2. $4x + 2x = 48$ $20x - 4x = 64$ $7x - 2x = 80$ $6x + 2 = 38 + 2x$

$X=24$ $x=8$ $x=\frac{1}{32}$ $x=16$ $X=9$

3. $3x + 6 = 41 - 2x$ $7x = 24\frac{3}{7}$ $6x + 9 = 29 + 2x$ $x = 36 - x$ $X=8$

$X=45$ $x=7$ $X=5$ $X=18$

4. $\frac{9}{7} = \frac{x}{21}$ $\frac{2}{5} = \frac{x}{15}$ $\frac{18}{4} = \frac{r}{6}$ $\frac{c}{4} = \frac{5}{10}$

5. $10^2 =$ $2^3 =$ $4^4 =$ $(5^2)^2 =$

Solve.

6. Triangle **ABC** is similar to triangle **XYZ**. What is the length of **XY**?

8 in.

7 in.

24 in.

Answer _____

7. The length and the width of a rectangle are in the ratio of 4 to 3. The perimeter is 1,400 in. Find the two dimensions.

Answer _____

Give the coordinates of each point.

8. A (____, ____)   E (____, ____)

   B (____, ____)   F (____, ____)

   C (____, ____)   G (____, ____)

   D (____, ____)

105

Unit 4   Review
Making Sure of Percent

Change to percent.

1. 0.17 = 17%   0.10 = 10% $\frac{1}{4}$ = 25%   0.625 = 62.5%   0.375 = 37.5%   1.5 = 1.5%   0.01 = 1%

(handwritten notes above: 62.5, 62½, 37.5 or 37½)

Change to decimals.

2. 15% = 0.15   0.1% = 0.1   25% = 0.25   20% = 0.20   12$\frac{1}{2}$% =   1% = 0.01   125% = 0.125

Write each percent as a decimal and as a fraction.

3. 5% = 0.05 $\frac{5}{10}$ 47% = 0.47 $\frac{47}{100}$

(handwritten: $\frac{5}{100}$, 1.00)

Find each number.

4. 20% of 25 = 25   30% of what number is 27?   90   What percent of 80 is 12?   150%

Solve these word problems.

5. Elizabeth bought an $18 shirt at a 25%-off sale. How much did she save? Answer $4.50	**6.** Mrs. White bought a $90 radio at 15% off. How much did she save? Answer $13.50	**7.** A store advertised all its goods at 20% off. Diane Barker bought a radio (marked $144) and a hair dryer (marked $18.95). How much did she pay? Answer 130.36
8. If we pay our electric bill by the tenth of the month, we can save 5%. If our bill is $60, how much should we pay after the discount? Answer $57	**9.** David Thaxton's monthly salary is $1,000 plus 8% commission on his total sales. Last month David sold $16,388. What was his salary? 2311.04 Answer $20,485.50	**10.** An airline is running an excursion at 25% savings. The regular fare is $60.80. How much less is the excursion fare? Answer $15.20
11. A store gets a discount of 5% for cash payment. How much will the store save by paying cash for a purchase of $18,650? Answer $9,325.0	**12.** The doctor has offered to settle all old accounts at 25% off. She collected $1,590. What was the total of the original accounts? 2120 Answer $6,360	**13.** At a sale a $975 piano was purchased for $585. What percent of the original amount was this? Answer 60%

Making Sure of Interest

Using the formula *I = prt*, find the interest on the following. Round to the nearest cent.

1. $400 at 4% for 1 year	$650 at 3% for 1 year	$250 at $3\frac{1}{2}$% for 2 years
1,600	1,950	17.50
2. $1,000 at $2\frac{1}{2}$% for 1 year	$325 at 3% for 6 months	$450 at $3\frac{1}{4}$% for 3 years
25	9.75 4.88	43.56
3. $1,250 at $2\frac{1}{4}$% for 1 year	$675 at 4% for $\frac{1}{4}$ year	$190 at $3\frac{1}{2}$% for 1 year
28.13	6.75	6.65

Solve.

4. Florence has $250 in the savings bank at 5%. How much simple interest will this amount earn in a year?

Answer _____

5. Madge's uncle gave her $100 for a birthday present. She put it in a bank at 4% interest, compounded semiannually. How much will this account amount to in a year's time? (Find the interest for 6 months; then find it on the new principal for the second 6 months.)

Answer _____

6. I put $1,200 in a savings account which pays 6% interest, compounded semiannually. How much will I have at the end of one year?

Answer _____

7. Jean invested in a $4,500 mortgage at 6% interest. The mortgage was paid off at the end of $3\frac{1}{2}$ years. How much interest was earned in that time?

Answer _____

8. Mr. Turner has $2,000 in a bank, drawing interest at 5%. He has a chance to invest it in a good mortgage which pays 12% interest. How much more interest per year will the mortgage earn?

Answer _____

9. Interest of $24 was paid on a 6% loan for 2 years. What was the amount of the loan?

Answer _____

107

Making Sure of Measurement

Change each measurement to a smaller unit.

1. 9 lb = __144__ oz $6\frac{1}{3}$ yd = _____ ft $4\frac{1}{2}$ qt = _____ pt

2. 28.2 L = __28200__ mL   7.8 cm = _____ mm   0.59 kg = _____ g

3. $2\frac{1}{2}$ ft = __30__ in. $3\frac{1}{4}$ lb = _____ oz   5 qt = _____ c

4. 0.6 kg = __6,000__ g   17.8 L = _____ mL   9.4 m = _____ cm

5. 3 yr = __156__ wk   4 hr = _____ sec   5 min = _____ sec

6. 2 wk = __14__ day   3 decades = _____ yr   63 wk = _____ day

Change each measurement to a larger unit.

7. 98 in. = __8 6__ ft   100 oz = _____ lb   6 qt = _____ gal

8. 7,400 mL = __7.4__ L   19.3 g = _____ kg   132 cm = _____ m

9. 5,000 lb = __$2\frac{1}{2}$__ T   8 ft = _____ yd   36 qt = _____ gal

10. 71 mm = __0.071__ cm   13.5 g = _____ kg   900 mL = _____ L

11. 60 yr = __6__ decade   72 hr = _____ day   540 sec = _____ min

12. 252 day = __36__ wk   730 day = _____ yr   156 hr = _____ day

Solve.

13. How many feet are there in a football field 100 yards long?

    Answer _____

14. How many grams are in a can of beans that weighs 0.26 kg?

    Answer _____

15. This year Lucille weighs 106 lb 13 oz. A year ago she weighed 101 lb 12 oz. How much has she gained in the year?

    Answer _____

16. Mark bought a 6 lb 8 oz block of cheese. How many 4-oz packages can he get from that block?

    Answer _____

17. A kilometer is 1,000 meters. A meter is equal to 100 centimeters. How many centimeters are there in a kilometer?

    Answer _____

18. Dwayne's property is 369.2 meters wide. If it is divided into four equal lots, how wide will each lot be?

    Answer _____

19. Burton had a roll of shelf paper 9 meters long. How many centimeters of paper did Burton have?

    Answer _____

20. One liter of medicine costs $234.50. What will be the cost of 1 mL of this medicine? (Round to the nearest cent.)

    Answer _____

Making Sure of Formulas

Using the appropriate formula from these given here, solve the following.

$P = 2l + 2w$ $A = lw$ $V = lwh$ $A = \frac{1}{2}bh$ $C = \pi d$ $A = \pi r^2$ $V = \pi r^2 h$ $c^2 = a^2 + b^2$

1. The meeting room measures 24 ft by 20 ft. If one gallon of paint will cover 320 sq ft, how many gallons will be needed to paint the ceiling of the room?

   Answer ___1.5___

2. Frank Jackson plans to erect fencing completely around his farm, which is 1,200 yd long and 800 yd wide. Without making allowance for gates, how much will the fencing cost at $0.40 per yd?

   Answer ___$16.00___

3. A triangular garden has a base of 140 ft and a height of 50 ft. How many square feet are there in the area of the garden?

   Answer ___3500___

4. Jerrie Reynolds had a cement sidewalk laid in front of her home at a cost of $0.95 per square foot. The walk was 4 feet wide and 75 feet long. How much did the walk cost?

   Answer ___$2.85___

5. A cord of wood is a pile 8 ft long, 4 ft wide, and 4 ft high. How many cubic ft are there in a cord of wood?

   Answer ___128___

6. There are 1.25 cu ft in a bushel. How many bushels can be stored in a bin that measures 5 ft by 8 ft by 12 ft?

   Answer _____

7. A circular flower garden has a diameter of 18 ft. Around the garden there is a circular sidewalk which is 2 ft wide. What is the length around the outer edge of the sidewalk? (Hint: Make a drawing.)

   Answer _____

8. Using the information from problem 7, find how many square feet are covered by the sidewalk alone.

   Answer _____

9. How many cubic feet capacity has a tank whose diameter is 14 ft and whose height is 40 ft? ($r = \frac{1}{2}d$)

   Answer _____

10. The legs of a right triangle are 9 and 12 meters long. What is the length of the hypotenuse?

    Answer _____

Making Sure of Graphs

Solve. Use the bar graph. Round answers to the nearest cent.

1. How much more is earned per week by workers in the transportation equipment industry than by workers in the lumber and wood products industry?

   Answer _____

2. Workers in the furniture and fixtures industry work about 39 hours per week. What is their average hourly wage? Round your answer to the nearest cent.

   Answer _____

3. How much lower is the average yearly salary for workers in iron and steel foundries than the salary for workers in the motor vehicles industry?

   Answer _____

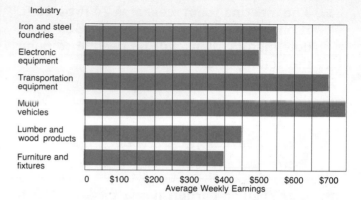

AVERAGE WEEKLY EARNINGS IN MANUFACTURING INDUSTRIES

Solve. Use the line graph.

4. In Year 5, about how much higher is Chris's salary than Pat's salary?

   Answer _____

5. What is the percent of increase in Pat's salary from Year 3 to Year 4?

   Answer _____

6. What is the percent of increase in Pat's salary from Year 4 to Year 5?

   Answer _____

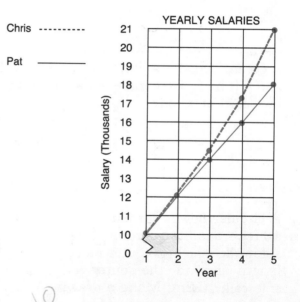

Solve. Use the circle graph.

If a circle is divided into 10 parts, each part is 10% of the whole. The circle graph at the right shows sources of Lola Dawson's income. The graph shows that 3 tenths or 30% was social security.

7. What percent of Lola's income was from her pension? _____

8. What percent of Lola's income was from investments? _____

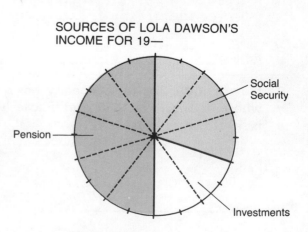

SOURCES OF LOLA DAWSON'S INCOME FOR 19—

Making Sure of Equations

Evaluate.

1. $20 - 17 =$ 3	$11 - 23 =$ -12	$-5 - (-3) =$ -2	$-9 - 2 =$ -7
2. $8(-6) =$ -48	$(-4)(-22) =$ +88	$\frac{-75}{3} =$ -25	$\frac{24}{-2} =$ -12
3. $3^2 =$ 9	$10^3 =$ 1,000	$\sqrt{25} =$ 5	$\sqrt{36} =$ 6
4. $8k + k =$ 9k	$12m - 3m -$ 9m	$7ab - 10ab =$ -3ab	$10n + (-3n) -$ 13n

Solve these equations for x.

5. $3x + 2x = 15$ X=3 $3x - 2 = 2x + 4 =$ 6x $3(x + 2) = 2(x + 4)$ 2

6. $3x - 2x = 15$ X=15 $5x + 2(x + 2) = 18$ X=2 $3x - 2 = x + 4 =$ 3

Set up an equation and solve these problems.

7. The length of a certain rectangle is twice its width. If the perimeter is 180 ft, what is the length of the rectangle?

Answer ___60 + 30___

8. The area of a certain square is 25 sq ft. How wide is the square?

Answer ___5 ft___

9. Two legs of a certain right triangle measure 3 ft and 4 ft. What is the perimeter of the triangle?

Answer ___12 ft___

10. I am thinking of a certain number. If I add 4 to the number and divide that sum by 11, the result is 2. What is the number?

Answer ___18___

11. A piece of wire 28 in. long is to be cut into two pieces so that one piece is 4 in. longer than the other. How long is each piece?

Answer ___10   18___

12. The sum of the three angles of any triangle is 180 degrees. In a certain isosceles triangle, each of the equal base angles contains 55 degrees. Find the size of the third angle.

Answer ___70___

13. Write the ordered pairs for the given points.

A (___, ___)

B (___, ___)

C (___, ___)

D (___, ___)

14. Plot and graph the ordered pairs.

A $(-1, 1)$
B $(3, 4)$
C $(-4, 4)$
D $(-3, -1)$

Making Sure of Ratio, Probability, and Proportion

Answer these problems involving ratio.

1. From the graph at the right, what is the ratio of the amount of cotton produced in this country to the amount produced in the rest of the world?

   Answer _____

2. What is the ratio of the amount produced in this country to that of the entire world?

   Answer _____

WORLD PRODUCTION OF COTTON

MILLIONS OF BALES

	10	20	30	40	50
World					
United States					
Rest of World					

Solve.

3. Juan, Ricardo, and Julio together have $450 and the amount each one has is in the ratio of 2 to 3 to 4. How much does each have?

   Juan _____ Ricardo _____ Julio _____

4. The angles of a triangle are in the ratio of 3:4:5. How many degrees are there in each angle? (Remember the sum of the angles is 180.)

   1st angle ____ 2nd angle ____ 3rd angle ____

5. Three classes in the Witt School contributed $360 to the Red Cross drive. Their contributions were in the ratio of 1 to 2 to 3. How much did each contribute?

   1st class ____ 2nd class ____ 3rd class ____

6. A tree casts a shadow of 10 feet at the same time that a fence post casts a shadow of 6 feet. The fence post is 4 feet tall. How tall is the tree?

   Answer _____

7. What is the probability of randomly drawing a white marble from a bag containing 3 red marbles, 2 blue marbles, and 9 white marbles?

   Answer _____

8. Alma can walk 10 miles in 3 hours. At that rate, how many miles can she walk in 4 hours?

   Answer _____

Solve each of the following proportions to find the value of the letter shown.

9. $x:21 = 2:3$	$5:1 = 30:x$	$12:x = 16:4$	$7:x = 28:8$
10. $35:14 = 50:x$	$4:x = 16:12$	$12:6 = 8:x$	$8:x = 32:12$
11. $\frac{21}{75} = \frac{x}{25}$	$\frac{3}{9} = \frac{x}{16}$	$\frac{27}{x} = \frac{15}{10}$	$\frac{20}{19} = \frac{30}{x}$

Basic Essentials of Mathematics

BOOK TWO

Mastery Test

Instructions

The test on pages 113-118 is designed to give you the opportunity to measure your progress at the end of this book. The teacher, of course, will want to know how well you have studied. But you will be most interested in proving to *yourself* how much you have learned, and in finding out whether you have acquired all the skills you will need.

There is no time limit for this review. Use all the time necessary to complete it. If you come to a problem which you cannot work, go on to the others. Then when you have finished them, go back to any you have skipped. Do your best.

The review is divided into five parts: Percent, Measurement, Formulas, Equations and Graphs, and Ratio and Proportion. When you have finished the review and it has been scored, enter your score in the chart on the right. This will enable you to determine exactly where there are any weaknesses and where you need to acquire more skill.

	PROBLEMS	PERFECT SCORE	MY SCORE
Percent			
Change to percent	1-3	3	_____
Change to decimals	4-6	3	_____
Change to fractions	7-9	3	_____
Finding percents	10-20	11	_____
Measurement			
Change to a smaller unit	21-32	12	_____
Change to a larger unit	33-44	12	_____
Computing measures	45-48	4	_____
Formulas			
Using formulas	49-55	7	_____
Applying formulas	56-64	9	_____
Equations and Graphs			
Solving equations	65-76	12	_____
Applying equations	77-80	4	_____
Graphing	81-88	8	_____
Ratio and Proportion			
Solving ratio and proportion	89-94	6	_____
Applying ratio and proportion	95-100	6	_____
TOTAL SCORE		100	_____

Percent

Change to percent.

1. $\frac{1}{2} =$ _____

2. $0.75 =$ _____

3. $0.375 =$ _____

Change to decimals.

4. $\frac{1}{4}\% =$ _____

5. $60\% =$ _____

6. $87\frac{1}{2}\% =$ _____

Change to fractions.

7. $0.75\% =$ _____

8. $40\% =$ _____

9. $62\frac{1}{2}\% =$ _____

Solve.

10. 14% of $300 =$ _____

11. 75% of $440 =$ _____

12. Taxes are often discounted if paid at an early date. Father received a 3% discount on his tax bill of $120. How much tax did he have to pay?

    Answer _____

13. The total enrollment in public elementary and secondary schools in 1900 was 75,995,000. The projected enrollment for a coming year is 295% of the 1900 count. What is the projected enrollment for that year?

    Answer _____

14. A grocer was allowed $2\frac{1}{2}\%$ discount for prompt payment of a bill. The discount was $10. How much was the original bill?

    Answer _____

15. Last year, a salesperson sold merchandise worth $160,500. Her commission of 6% amounted to how much?

    Answer _____

16. Last month the utility bill was $98.00. This month shows an increase of $12\frac{1}{2}\%$. How much is this month's bill?

    Answer _____

17. The distance from New York to San Francisco is 5,400 miles by way of the Panama Canal. This is 60% of the distance by way of the Strait of Magellan. How far is it by way of the Strait of Magellan?

    Answer _____

18. Mr. Evans, the grocer, borrowed $1,250 at 6% for 6 months. How much interest would he owe on the loan?

    Answer _____

19. Dan's bank pays 5% interest. How much interest will he have on his savings account of $360 at the end of a year?

    Answer _____

20. To help build some houses, Ms. Raymond borrowed $92,000 from the bank and agreed to pay it back in two and one-half years, with interest at 10%. How much interest will she pay?

    Answer _____

Measurement

Change each measurement to the smaller unit.

21. 3 mi = _____ yd

22. 4 lb = _____ oz

23. 17 m = _____ mm

24. $2\frac{1}{4}$ ft = _____ in.

25. $3\frac{3}{16}$ gal = _____ c

26. 5.7 km = _____ m

27. $1\frac{1}{2}$ T = _____ lb

28. 12 qt = _____ pt

29. 0.025 kg = _____ g

30. 9 wk = _____ days

31. 3 hr = _____ min

32. 2 min = _____ sec

Change each measurement to the larger unit.

33. 45 in. = _____ ft

34. 3,500 lb = _____ T

35. 4,139 m = _____ km

36. 66 ft = _____ yd

37. 19 c = _____ gal

38. 806 cm = _____ m

39. 29 oz = _____ lb

40. 85 qt = _____ gal

41. 152 g = _____ kg

42. 35 days = _____ wk

43. 96 hr = _____ days

44. 180 sec = _____ min

Find each answer.

45.
$$\begin{array}{r} 8 \text{ lb} \quad 9 \text{ oz} \\ + 1 \text{ lb} \quad 1\,2 \text{ oz} \\ \hline \end{array}$$

Answer _____

46.
$$\begin{array}{r} 6 \text{ gal} \quad 1 \text{ qt} \\ - 3 \text{ gal} \quad 3 \text{ qt} \\ \hline \end{array}$$

Answer _____

47. $3\overline{)8 \text{ min } 6 \text{ sec}}$

Answer _____

48.
$$\begin{array}{r} 3 \text{ yd} \quad 2 \text{ ft} \\ \times \qquad\quad 5 \\ \hline \end{array}$$

Answer _____

Solve.

49. $P = 2l + 2w$
$l = 400$ ft
$w = 300$ ft

$P =$ _____

50. $A = lw$
$l = 60$ m
$w = 50$ m

$A =$ _____

51. $V = lwh$
$l = 30$ in.
$w = 20$ in.
$h = 10$ in.

$V =$ _____

52. $C = \pi d$
$\pi = \frac{22}{7}$
$d = 14$ ft

$C =$ _____

53. $A = \pi r^2$
$\pi = \frac{22}{7}$
$r = 7$ m

$A =$ _____

54. $V = \pi r^2 h$
$\pi = \frac{22}{7}$
$r = 7$ cm
$h = 10$ cm

$V =$ _____

55. $I = prt$
$p = \$500$
$r = 4\%$
$t = 24$ mo

$I =$ _____

Use the correct formula to solve.

56. From the formula $A = lw$, write the formula for finding w.

Answer _____

57. What is the perimeter of the rectangle whose length is 60 meters and whose width is one-half the length?

Answer _____

58. How many feet of wire will be required to enclose a triangular - shaped garden whose sides are 150 ft, 140 ft, and 200 ft?

Answer _____

59. The new gymnasium floor measures 90 ft by 150 ft. How many square feet are there in this floor area?

Answer _____

60. How much did Monica pay at \$9 per square yard for a rug which measured 9 ft by 12 ft?

Answer _____

61. A cord of wood is a pile 8 ft long, 4 ft wide, with a volume of 128 cubic ft. What is the height of the pile?

Answer _____

62. How many cubic yards of dirt will be required to fill a hole 27 ft by 30 ft by 10 ft?

Answer _____

63. A coal car has inside dimensions of 40 ft by 6 ft and is filled level to a 5-ft depth. Allowing 40 cu ft to the ton, how many tons of coal are there in this car?

Answer _____

64. A water standpipe is 60 feet high and has a diameter of 14 feet. What is the volume of the standpipe (in cubic feet)? Use $\pi = \frac{22}{7}$.

Answer _____

116

Equations and Graphs

65. $5x + 3 = 18$

$x =$ _____

66. $5x - 1 = 4x + 3$

$x =$ _____

67. $5(x + 4) - 4(x + 6)$

$x =$ _____

68. $x - 15 = 7$

$x =$ _____

69. $x + 5 = 19$

$x =$ _____

70. $7x = 42$

$x =$ _____

71. $6x + 1 = 61$

$x =$ _____

72. $3x + 5x = 72$

$x =$ _____

73. $10x - 4x = 36$

$x =$ _____

74. $6x + 9 = 29 + 2x$

$x =$ _____

75. $6 = 2x - 20$

$x =$ _____

76. $10x = 5x + 15$

$x =$ _____

Solve using an equation.

77. Together Yoko and Teresa have $525. Yoko has $25 more than Teresa. How much does each have?

Answer _____

78. The sum of Father's age and my age is 65 years. Father is 25 years older than I am. How old is each?

Answer _____

79. The perimeter of a rectangular field is 60 meters. The length is twice the width. Find the length and width.

Answer _____

80. The sum of three numbers is 175. The second number is twice the first, and the third is twice the second. What are the numbers?

Answer ____ ____ _____

Write the ordered pairs for the given points.

81. A (__, __) **82.** B (__, __)

83. C (__, __) **84.** D (__, __)

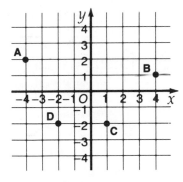

Plot and label the ordered pairs.

85. A $(-1, 4)$ **86.** B $(3, 3)$

87. C $(4, -2)$ **88.** D $(-4, -1)$

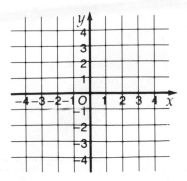

Ratio and Proportion

Solve.

89. $30:x = 2:3$

$x =$ _____

90. $\frac{x}{6} = \frac{5}{3}$

$x =$ _____

91. $10:50 = x:25$

$x =$ _____

92. $\frac{4}{x} = \frac{12}{15}$

$x =$ _____

93. $x:15 = 3:5$

$x =$ _____

94. $\frac{5}{6} = \frac{x}{24}$

$x =$ _____

Solve using ratio and proportion.

95. Chi has $3.00, and Sun has $2.00. What is the ratio of Chi's money to that of Sun?

Answer _____

96. The perimeter of a triangle is 81 meters. The three sides are in the ratio of 2 to 3 to 4. Find the length of each side.

Answer _____

97. Diva, Clara, and Lucia together have $45. The ratio of their amounts is 4 to 5 to 6. How much does each one have?

Answer _____

98. If a car uses 8 gallons of gas to go 160 miles, how much gas is needed to go 200 miles?

Answer _____

99. Rosa earns $120 in 2 days. How much does she earn in 5 days?

Answer _____

100. A telephone pole casts a shadow of 30 feet at the same time a fence post casts a shadow of 3 feet. The fence post is 6 feet tall. How tall is the telephone pole?

Answer _____

118

Answer Key

Unit 1

Page 5
1. $32
2. $6.40
3. $50
4. $37.20
5. $25
6. $6.25
7. $18
8. $28
9. $11.25
10. $999

Page 6
1. 0.20 0.15 0.10 0.25 0.50 0.75
2. 0.30 0.40 0.12 0.18 0.22 0.27
3. 0.145 0.176 0.601 $0.33\frac{1}{2}$ $0.45\frac{1}{4}$ $0.35\frac{2}{5}$
   or or or
   0.335 0.4525 0.354
4. 0.90 0.80 0.70 0.65 0.95 1.00
5. 0.01 0.07 0.05 0.09 0.08 0.03
6. 10% 20% 30% 40% 50% 60%
7. 15% 25% 35% 45% 65% 75%
8. 12% 27% 43% 67% 90% 100%
9. 1% 5% 9% 7% 3% 8%
10. 10% 20% 30% 70% 90% 50%
11. 12.5% 37.5% 62.5% 87.5% 8.5% 1.5%
   or or or or or or
   $12\frac{1}{2}$% $37\frac{1}{2}$% $62\frac{1}{2}$% $87\frac{1}{2}$% $8\frac{1}{2}$% $1\frac{1}{2}$%

Page 7
1. 50% 25% 75% 20% 40%
2. 60% 80% 30% 70% 90%
3. 37.5% 62.5% 87.5% 3.125% 15.625%
   or or or or or
   $37\frac{1}{2}$% $62\frac{1}{2}$% $87\frac{1}{2}$% $3\frac{1}{8}$% $15\frac{5}{8}$%
4. 6.25% 18.75% 31.25% 43.75% 68.75%
   or or or or or
   $6\frac{1}{4}$% $18\frac{3}{4}$% $31\frac{1}{4}$% $43\frac{3}{4}$% $68\frac{3}{4}$%
5. $33\frac{1}{3}$% $11\frac{1}{9}$% $66\frac{2}{3}$%

Page 8
1. $\frac{1}{2}$ 0.50 50%
2. $\frac{1}{4}$ 0.25 25%
3. $\frac{3}{4}$ 0.75 75%
4. $\frac{2}{5}$ 0.40 40%
5. $\frac{3}{5}$ 0.60 60%
6. $\frac{1}{10}$ 0.10 10%
7. $\frac{7}{10}$ 0.70 70%
8. $\frac{1}{4}$ 0.25 25%
9. $\frac{3}{10}$ 0.30 30%
10. $\frac{5}{8}$ 0.625 $62\frac{1}{2}$% or 62.5%
11. $\frac{9}{20}$ 0.45 45%
12. $\frac{3}{20}$ 0.15 15%
13. $\frac{18}{25}$ 0.72 72%
14. $\frac{7}{50}$ 0.14 14%
15. $\frac{1}{25}$ 0.04 4%

16. $\frac{1}{20}$ 0.05 5%
17. $\frac{3}{20}$ 0.15 15%
18. $\frac{1}{5}$ 0.20 20%
19. $\frac{1}{0}$ 0.125 $12\frac{1}{2}$% or 12.5%
20. $\frac{3}{8}$ 0.375 $37\frac{1}{2}$% or 37.5%
21. $\frac{1}{3}$ $0.33\frac{1}{3}$ $33\frac{1}{3}$%
22. $\frac{2}{3}$ $0.66\frac{2}{3}$ $66\frac{2}{3}$%
23. $\frac{4}{5}$ 0.80 80%
24. $\frac{1}{12}$ $0.08\frac{1}{3}$ $8\frac{1}{3}$%
25. $\frac{5}{8}$ 0.625 $62\frac{1}{2}$%
26. $\frac{7}{8}$ 0.875 $87\frac{1}{2}$%
27. $\frac{4}{5}$ 0.80 80%
28. $\frac{9}{10}$ 0.90 90%
29. $\frac{7}{10}$ 0.70 70%

Page 9
1. 35%
2. 75%
3. 0.0725
4. $\frac{9}{25}$
5. 9.75% or $9\frac{3}{4}$%
6. 17%
7. 0.055
8. 40%
9. 0.01
10. 0.995
11. 80%
12. 12%

Page 10
1. 3 4 6 8 10
2. 1.50 = 1.5 1.25 1.75 1.99 5.00 = 5
3. 2.25 2.50 = 2.5 1.125 1.375 1.625
4. 3.10 = 3.1 3.375 3.8775 1.332 1.664
5. 1.01 1.60 = 1.6 1.05 1.20 = 1.2 2.02
6. 2.05 9.999 4.50 = 4.5 8.07 3.20 = 3.2
7. 6.09 7.255 11.00 = 11 1.013 1.7525
8. 2.10 = 2.1
9. 1.52
10. 3.33
11. 1.35
12. 1.16
13. 3.00 = 3

Page 11
1. 0.005 0.0033 0.0025 0.006
2. 0.004 0.0041 0.0005 0.0075
3. 0.0025 0.004 0.006 0.008 0.00125
4. 0.00375 0.00625 0.00875 0.001 0.003
5. 0.007 0.009 0.003125 0.004375 0.0005
6. 0.009
7. 0.0064
8. 0.0055
9. 0.45%
10. $\frac{56}{100}$% or 0.56%
11. $\frac{999}{1000}$, 99.9% or $99\frac{9}{10}$%, 0.999

Page 12
1. 45   180   30   90   21
2. 96   60   18   12   11.25
3. 30   180   20   42   96
4. 250   54.45   100   60   70
5. 72   245   576   300   36
6. 45   43   67.5   18   9
7. 39   150   100   27   42.15

Page 13
1. $3,250   2. $6,000
3. $7,200   4. $3,600
5. $4,800   6. $1,920
7. 200 cars   8. 30 pounds
9. 6.45 pounds   10. 0.344 pound

Page 14
1. 4 $\frac{2}{3}$ 3
2. 34 $1\frac{1}{3}$ 9
3. 33   3   12
4. $1\frac{1}{5}$ $4\frac{2}{3}$ 15
5. $3\frac{3}{8}$ 8   5
6. $2\frac{1}{4}$ 14 $2\frac{4}{5}$
7. 9   36 $2\frac{39}{40}$

Page 15
1. $10   2. $9   3. $26
4. $7   5. $20   6. $30
7. $40   8. $24.50   9. $7.50
10. $18.38
11. $18
12. $60

Page 16
1. $2   2. $20   3. $5
4. $14   5. $30   6. $18
7. $20   8. $112.50   9. $40.50
10. $490   11. $89.60   12. $506

Page 17
1. $312.12   2. $418.20
3. $525.31   4. $331.15

Page 18
1. 491,520 people   2. $294,000
3. $106.04   4. $27,000
5. 1,845 people   6. 80 bushels
7. $9,275   8. 360 pounds

Page 19
1. $2,625   2. 14 pounds
3. 13.6 quarts   4. $85.75
5. $68   6. 7.2 inches
7. 2,160 people   8. $20,825

Page 20
1. $11,532
2. $83,220
3. $657.50
4. $3,798
5. $738
6. $30,080
7. $1,178

Page 21
1. 68   125   40
2. 80   200   57,000
3. 55   200   13,000
4. 730   160   40
5. 200   7.5   20
6. 25   300   35
7. 200   90   900
8. 80   540   50

Page 22
1. 720 families   2. $40,000
3. $20   4. 125 acres
5. $75,000   6. 360 acres
7. $36,000   8. 16 pounds

Page 23
1. $12,400   2. $30
3. $18   4. $80
5. $1,000   6. $1,000
7. $10,000   8. $2,000

Page 24
1. $500   2. $2,160
3. 200 acres   4. 2,400 people
5. $6,600   6. 13,000 miles
7. $450   8. 16 square yards
9. 1,250 people   10. 200 pounds

Page 25
1. 20%   60%   7.5% or $7\frac{1}{2}$%
2. 15%   30%   4%
3. 45%   15%   85%
4. 67%   80%   80%
5. $10\frac{2}{3}$%   37.5% or $37\frac{1}{2}$%   75%
6. 12%   20%   40%
7. 12.5% or $12\frac{1}{2}$%   10%   87.5% or $87\frac{1}{2}$%
8. 25% $66\frac{2}{3}$%   35%

Page 26
1. 20%   2. 40%
3. 75%   4. 25%
5. 10%   6. 25%
7. $33\frac{1}{3}$%   8. 12.5% or $12\frac{1}{2}$%

Page 27
1. 33   22. 9
2. 32   23. 24
3. 25   24. 40
4. 108   25. 21
5. 1   26. 25
6. 10   27. 0.66
7. 200   28. 200
8. 80   29. 10
9. 35   30. 20
10. 8.5   31. 250
11. 80   32. 17
12. 13.5   33. 0.17
13. 0.5   34. 40
14. 505
15. 5   35. 14%   36. $194.22
16. 2
17. 2
18. 9
19. 160
20. $10.75
21. 9

120

Page 28
1. $0.45
2. Evelyn, $680
3. 8%
4. 6 hours

Page 29
1. 16 problems
2. $49.60
3. $18
4. $40
5. $25
6. $37.50
7. $55.60
8. 5,000 acres
9. $11.25
10. $315
11. $7.54
12. $119

Page 30
1. 0.004, $\frac{1}{250}$ 0.355, $\frac{71}{200}$ 0.11, $\frac{11}{100}$ 2.5, $2\frac{1}{2}$
2. 87.5% or $87\frac{1}{2}$%   345%   0.3%   16%
3. 40.5 or $40\frac{1}{2}$ 1.80 = 1.8 or $1\frac{4}{5}$ 2.45 or $2\frac{9}{20}$
4. 16%   200%   68
5. $112
6. $21.29
7. $20
8. $950
9. $4.50
10. $6.42

Unit 2
Page 31
1. 260   210   884   780
2. 1,825   10,800   1,728   17,520
3. 30   2 $1\frac{1}{2}$ 3
4. 2   12 $3\frac{1}{2}$ $5\frac{1}{2}$
5. $2\frac{1}{2}$ hours
6. 2,700 seconds

Page 32
1. $15\frac{1}{2}$ 7,920   88   126,720
2. 45   63   7,040   144
3. $4\frac{1}{4}$ $12\frac{2}{3}$ $\frac{1}{2}$ $\frac{1}{4}$
4. $2\frac{1}{2}$ $6\frac{1}{6}$ 3 $\frac{2}{3}$
5. $\frac{2}{3}$ yard
6. 880 yards

Page 33
1. 52   5,000   96   32,000
2. 8,000   22   67   1,000
3. $3\frac{5}{16}$ $3\frac{1}{2}$ 5 $\frac{2}{5}$
4. $4\frac{1}{2}$ $2\frac{1}{4}$ $1\frac{1}{5}$ 1
5. 228 ounces
6. 11,000 pounds

Page 34
1. 17   9   26   64
2. 33   36   24   240
3. $1\frac{3}{4}$ $2\frac{5}{8}$ $5\frac{1}{2}$ 4
4. 6   7 $3\frac{3}{8}$ $2\frac{1}{2}$
5. 64 jars
6. $1\frac{1}{2}$ pints

Page 35
1. 1 lb 14 oz   1 ft 6 in.   11 gal 4 qt
2. 8 gal 1 qt   14 ft 2 in.   11 yd
3. 1 yd 2 ft   1 gal 3 qt   13 lb 12 oz
4. 73 lb 8 oz   39 ft 8 in.   50 T
5. 1 gal 3 qt   1 lb 5 oz   1 yd 1 ft

Page 36
1. 18 ft 2 in.   10 lb   12 hr
2. 6 yd 1 ft   8 T   10 mi 140 yd
3. 3 yd 1 ft   1 qt 1 pt   3 lb 12 oz
4. 3 gal 1 qt   2 T 1,400 lb   10 ft 5 in.
5. 24 lb 12 oz   27 T   18 ft 9 in.
6. 15 gal   8 yd   26 qt
7. 2 min 42 sec   1 T 700 lb   1 yd 2 ft
8. 1 qt 1 pt   1 day 8 hr   1 lb 6 oz
9. 17 yr 10 mo
10. 4 yd 7 in.

Page 37
1. 14,500   725   180   6,500
2. 340   21,000   900   290
3. 4.8   0.796   0.061   0.657
4. 8.542   3.128   9.30 = 9.3   7.5
5. 1,500 centimeters
6. 32.5 kilometers

Page 38
1. 32,000   7   1,800   526,000
2. 490   825,000   6,783   80
3. 0.0128   0.009   0.137   0.02325
4. 5.268   0.025   0.0049   0.789
5. 25,000 grams
6. 0.246 kilograms

Page 39
1. 27,000   5,300   7,450   500
2. 825   2,000   39,600   2,600
3. 3.096   6   0.4125   0.528
4. 0.058   0.798   0.0192   0.00634
5. 4,500 milliliters
6. 0.250 liters

Page 40
1. $\frac{1}{4}$ $1\frac{13}{36}$
2. $26\frac{2}{3}$ $2\frac{2}{3}$
3. 0.064 $\frac{3}{10}$
4. $1\frac{1}{4}$ 8,609
5. 912   1,000
6. 0.001   63,360
7. 65,300 $\frac{1}{16}$
8. 41 min   5 hr 50 min   13 days 6 hr
9. 2 ft 9 in.   1 yd 2 ft   1 T 1,400 lb
10. 25 hr   27 bu   14 T 800 lb
11. 3 hr 20 min   1 pk 4 qt   1 day 5 hr
12. 1 T 1,100 lb
13. $15.75

Page 41
1. 6.9 meters
2. 800 lb
3. 31 yr 9 mo
4. 26 yr 9 mo
5. $4.56
6. $344
7. $24.75
8. 1 yd 9 in.
9. 1 T 1,300 lb
10. 185 lb 12 oz
11. 9 in.
12. 50 miles per hour

Page 42
1. 15 blocks
2. 9 blocks
3. 18 blocks
4. 18 blocks
5. 30 blocks
6. 51 blocks
7. 175 miles
8. 30 ft
9. 1,800 miles
10. 10 inches

Page 43
1. 70 meters
2. 0.85 kilometer
3. 232 centimeters
4. 558 centimeters
5. 5.2 kilometers
6. 123 millimeters

Page 44
1. 55.2 square meters
2. 384 square inches
3. 2,480 square centimeters
4. 300 square meters
5. 150 square meters
6. 2,268 square feet
7. 80 square inches
8. 13.2 square meters

Page 45
1. 4 square units · 9 square units · 36 square units
   8 units · 12 units · 24 units
2. 144 square feet
3. 60 feet

Page 46
1. 1
2. 1
3. 4
4. 4
5. 8
6. 8
7. 32
8. 32 cu in., yes
9. 1,000
   6,000
   15,232
   15
   2,500
10. 1,440
   1,920
   6,336
   281.6
   1,020

Page 47
1. parallel · intersecting
2. perpendicular · vertical
3. line · angle
4. acute · 4 right angles
5. 3 acute angles · obtuse

Page 48
1. isosceles · scalene · equilateral · scalene
2. right · obtuse · equiangular · acute

Page 49
1. $16\frac{1}{2}$ ft
2. 13.5 ft
3. Yes
4. No
5. $13\frac{1}{2}$ ft
6. 490 ft
7. 22 ft
8. 210 centimeters
9. 164 meters
10. 22 kilometers

Page 50
1. 160 sq ft
2. 375 square yards
3. 43,200 sq ft
4. 150 square feet

Page 51
1. 30°
2. 80°, 40°
3. 72°
4. 80°
5. 60°
6. 45°
7. 50°, 100°

Page 52
1. 8; 4,096
2. 9; 6,561
3. 10; 10,000
4. 12; 144
5. 20; 400
6. 25
7. 5
8. 10 feet
9. 20 feet

Page 53
1. 22 meters
2. 550 feet
3. 44 millimeters
4. 88 ft
5. 220 centimeters
6. 19.79 or $19\frac{4}{5}$ meters

Page 54
1. 38.5 or $38\frac{1}{2}$ square meters
2. 616 square meters
3. 38.5 or $38\frac{1}{2}$ square meters
4. 13.9 or $13\frac{43}{50}$ square meters
5. 7 gallons
6. 55.42 or $55\frac{11}{25}$ square centimeters

Page 55
1. 720 revolutions
2. $56\frac{4}{7}$ or 56.57 in.
3. 616 square meters
4. $9\frac{5}{8}$ or 9.625 sq in.
5. 120°, 240°
6. 214 sq ft
7. 28; 88; 616 · 21; 132; 1,386 · 42; 132; 1,386
   7; 44; 154 · 7; 22; 38.5 · 10.5; 66; 346.5
   14; 44; 154 · 3.5; 22; 38.5 · 1.75; 11; 9.625

Page 56
1. 9,240 cubic feet
2. $6,187\frac{1}{2}$ or 6,187.5 cubic feet
3. 138,600 gallons
4. 3,080 cubic feet
5. 4 times
6. $339\frac{3}{7}$ or 339.43 cu in.

Page 57
1. 164 ft or $54\frac{2}{3}$ yards
2. 111 ft
3. 6,880 yards
4. $\frac{7}{10}$ mile; 3,696 feet
5. 920 ft
6. 2,300 ft
7. $3.60
8. 123 ft

Page 58
1. 144
8. 0.025 or $\frac{1}{40}$ acre
2. 3 square feet
9. $233\frac{1}{3}$ square yards
3. 9
10. $675
4. 3 square yards
11. $294
5. 9
12. $43.20
6. 36 square feet
13. 100 square feet
7. 4,840
14. $2,520

Page 59
1. 40 cubic yards
2. 20 cords
3. $310
4. 60,000 lb
5. 320 loads
6. 8 cubic yards
7. 24 gallons
8. 5,808 cu ft
9. $6,720

Page 60
1. 960 bushels
2. 38,400 bushels
3. 60,750 gallons
4. 20 gallons
5. 30 tons
6. 5 gallons
7. 320 bushels
8. $2\frac{1}{2}$ gallons

Page 61

1. 8	12	45	8. 15	13	84		
2. 5	12	90	9. 3	24	91		
3. 6	7	60	10. 2	60	120		
4. 5	3	40	11. 25	44	600		
5. 8	25	96	12. 2	56	11		
6. 4	12	84	13. 5	100	12		
7. 15	10	72	14. 3	75	125		

122

Page 62
1. 12% 2. 5%
3. 2 years 4. 2 years
5. 2 years 6. $200
7. $100 8. $120

Page 63
1. 20 meters 2. 100 yards
3. 50 feet 4. 150 feet
5. 20 in. 6. 500 feet
7. 150 feet 8. 200 feet
9. 10, 40, 20, 30, 25
10. 10, 20, 20, 30, 50

Page 64
1. 4 inches 2. 15 inches
3. 11 in. 4. 10 inches
5. 6 feet 6. 12 feet
7. 20 feet 8. 10, 5; 10, 4; 10

Page 65
1. 300 miles per hour 2. 360 miles per hour
3. $1\frac{3}{4}$ hours 4. 8 hours
5. 375 miles 6. 150 miles
7. 6 hours 8. 436 miles

Page 66
1. 19.2% 2. 25%
3. 48% 4. 19.2%
5. 19.2% 6. 36%

Page 67
1. 800 cups 2. chocolate, peach
3. peach 4. 125 more cups
5. 50 fewer cups 6. 200 more cups

Page 68
1. 6 years 2. hundred thousands
3. $100,000 4. $100,000
5. $500,000 6. 1982-1983; $200,000

Page 69
1. $9,800; $4,200; $7,000; $5,600; $1,400
2. 60%; $16,800
3. 75%
4. $2,800
5. Parks and Sanitation
6. Inspection
7. $2,016,000
8. 37.83%
9. $1,716,000

Page 70
1. 2,636 KWH 2. 744 CCF 3. 264 KWH; $21.12
4. 2,245 miles
5. 3,265 miles
6. 2,115 miles
7. 670 miles
8. 1,749 miles
9. 58.3 gallons

Page 71
1. 40; 96 2. 960 3. 216
   54;180 3,000 300
   84;440 4,800 330

4. 14;44 5. 28;616 6. 7;1,540
   7;44 3.5;38.5 42;16,632
   10.5;66 21;1,386 14;12,320

7. 5,280 27 9,200
8. 24 $3\frac{3}{4}$ 4.139
9. 10 lb 5 oz 2 gal 2 qt 18 yd 1 ft

Page 72
1. 27.5 meters 2. 120 feet
3. 2,400 square feet 4. 135 square feet

Page 73
1. $2,300 2. $750
3. $360 4. 35 square feet
5. 60 cubic feet 6. 400 cubic yards
7. 200 cubic yards 8. 4,620 cubic feet
9. 864 cubic feet 10. 6,480 cubic ft

Page 74
1. 9,680 42 3,000
2. 17,000 5,700 25
3. $7\frac{1}{2}$ 22 $1\frac{13}{16}$
4. 3.5 0.019 0.850 = 0.85
5. 14 lb 1 oz 3 gal 3 qt 57 yd
6. 1 T 500 lb 3 hr 45 min 38 days
7. 30 blocks 8. 14 square yards
9. 30 inches 10. 2,304 cubic inches
11. 88 feet 12. 3.45 square meters

Unit 3
Page 75

1. -7 9 $\frac{1}{10}$ 14

2. 5x 2xy $\frac{1}{3}a$ 12z

3. 5 4 1.5 -42 1,250 1.5 -9 $\frac{2}{3}$

4. $\frac{1}{2}$ $-\frac{3}{4}$ $\frac{1}{5}$ $\frac{3}{2}$ $-\frac{2}{5}$ $\frac{1}{4}$ -1 $\frac{3}{5}$

Page 76
1. 11 15 80 52
2. 4 4 5 2
3. 16 16 21 18
4. 43 70 27 27
5. 42 90 90 34
6. 6 6 12 6

Page 77
1. -30 89 -76
2. 18 -13 -15
3. -4 -3 14
4. -1 3 0
5. -10 -43 -48
6. 36 42 -42
7. -14 -6 8
8. 48 91 -115
9. -6 12 -24
10. -14 -18 16
11. -3 -3 8

Page 78

Variable can be any letter.

1. $5 + x$ $= 25;$ $5 + x = 25$
2. $\frac{2x}{7}$ $= 14;$ $\frac{2x}{7} = 14$
3. $x, x + 14$ $= 32;$ $x + x + 14 = 32$
4. $x, 7x$ $= 320;$ $x + 7x = 320$

Page 79

1. 8   30   5   4
2. 4   7   16   25
3. 300   11   15   66
4. 9   0   1   12
5. $\frac{1}{7}$ 44   80   505
6. $\frac{1}{2}$ $1\frac{1}{2}$ $4\frac{1}{3}$ $3\frac{1}{4}$
7. $x + 4 = 10$ $x = 6$

Page 80

1. -20 6   24   5
2. 20   5 -19 -2
3. 6 -55 23   5
4. $3x + 4x = 42$ $x = 6$

Page 81

1. 7 -4 4   3
2. -1 3   4   1
3. 9   4 -2 2
4. $4x + 100 = 2,500$ $x = \$600$

Page 82

1. -5 4 -1
2. 5   0 -4
3. 2 -44 1
4. $2x + 5 = 3x - 5$ $x = 10$

Page 83

1. 4   8 -4
2. 5   4   1
3. 6 -5 4
4. 5   4   10
5. 6   3 -5

Page 84

1. $\$9.60$ $\$3.20$ 2. 48
3. $\$35$ $\$7$ 4. $\$35$ $\$20$
5. 50 envelopes, 30 envelopes   6. 75 sq ft, 125 sq ft
7. 54 yr old, 18 yr old   8. 58 yr old, 20 yr old

Page 85

1. P (3, 4)   I (0, 2)
2. E (2, -2)   B (-6, -7)
3. D (7, 0)   G (0, -5)
4. H (-2, 7)   K (-4, 0)
5. J (-4, -2)   A (3, 7)
6. C (8, -6)   F (-7, 2)

Page 86

1.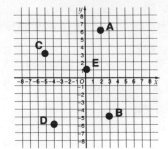

2.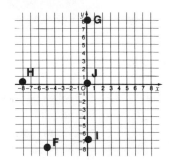

Page 87

1. 14   5   16
2. $a + 8$ $-27a$ $2a + 7b$
3. 18 -2 45
4. 3 $\frac{-8}{9}$ 4
5. -5 $\frac{5}{3} = 1\frac{2}{3}$ 4
6. 8 -5 -1
7. $\frac{1}{3}$ 2 -3
8. 7   3   3
9.

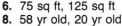

Page 88

1. $12:36$ or $\frac{12}{36} = \frac{1}{3}$ $24:168$ or $\frac{24}{168} = \frac{1}{7}$
2. $2:4$ or $\frac{2}{4} = \frac{1}{2}$ $3:6$ or $\frac{3}{6} = \frac{1}{2}$
3. $25:100$ or $\frac{25}{100} = \frac{1}{4}$ $8:10$ or $\frac{8}{10} = \frac{4}{5}$
4. $50:10$ or $\frac{50}{10} = \frac{5}{1}$ $10:8$ or $\frac{10}{8} = \frac{5}{4}$
5. $17:20$ or $\frac{17}{20}$ $5:3$ or $\frac{5}{3}$
6. $31:31$ or $\frac{31}{31} = \frac{1}{1}$ $8:16$ or $\frac{8}{16} = \frac{1}{2}$
7. $60:30$ or $\frac{60}{30} = \frac{2}{1}$ $8:8$ or $\frac{8}{8} = \frac{1}{1}$
8. $10:4$ or $\frac{10}{4} = \frac{5}{2}$ $2:4$ or $\frac{2}{4} = \frac{1}{2}$

124

Page 89
1. 30 feet, 10 feet
2. 8 in., 16 in., 24 in.
3. 200 meters, 160 meters, 120 meters
4. Tom, $0.40; Dick, $0.80; Harry, $1.20
5. $0.60, $0.90, $1.20
6. $21, $28, $35
7. $12, $16, $20
8. 600, 900, 2,100

Page 90
1. $\frac{5}{15} = \frac{1}{3}$
2. $\frac{1}{6}$
3. $\frac{1}{3}$
4. $\frac{1}{4}$
5. $\frac{8}{16} = \frac{1}{2}$
6. $\frac{2}{6} = \frac{1}{3}$

Page 91
1. $2.50:$2.00 = $5:$4 or $\frac{5}{4}$
2. $2.00:$2.50 = $4:$5 or $\frac{4}{5}$
3. 6,000:40,000 = 6:40 or $\frac{6}{40} = \frac{3}{20}$
4. $\frac{1}{2}$
5. $\frac{4}{9}$
6. $1.20, $2.40, $4.80
7. $\frac{1}{2}$
8. $\frac{1}{3}$
9. $\frac{6}{19}$
10. $\frac{1}{6}$
11. 5:125 or $\frac{5}{125} = \frac{1}{25}$
12. $\frac{3}{6}$ or $\frac{1}{2}$
13. 30, 60, 90
14. $6.28

Page 92
1. 72 in.:64 in. or $\frac{72}{64} = \frac{9}{8}$
2. 50 years old
3. 64 in.:72 in. or $\frac{64}{72} = \frac{8}{9}$
4. 8:2 = 4:1 or $\frac{4}{1}$
5. 3,500 adults
6. 40 ft
7. $\frac{1}{8}$
8. $\frac{4}{8} = \frac{1}{2}$
9. $\frac{5}{10} - \frac{1}{2}$
10. $\frac{2}{10} = \frac{1}{5}$
11. $\frac{3}{20}$
12. $\frac{10}{20} = \frac{1}{2}$

Page 93
1. 30   24   3   16
2. 12   21   2   14
3. 2   16   12   4
4. 2   12   2   15

Page 94
1. 3 $1\frac{1}{5}$ 2   6
2. 3   7   10   10
3. 10   6 $1\frac{1}{5}$ 6
4. 3   15   12   13

Page 95
1. 20 miles
2. 80.8 seconds
3. 9 seconds
4. $75
5. $5.45
6. $90.75
7. 30 miles
8. 20.24 seconds

Page 96
1. 8 times
2. 10 cu yd of sand, 15 cu yd of gravel
3. 4 times
4. 6.25 times
5. $3\frac{3}{8}$ times
6. 24 inches
7. 8 times
8. 26 pounds of tin, 39 pounds of lead

Page 97
1. 21 in.
2. $3\frac{2}{3}$ ft
3. 8 m
4. 8 cm
5. $\frac{1}{2}$ yd
6. 2 ft

Page 98
1. AB   AC
2. DF   DF
3. 50 feet
4. 555 feet
5. 40 feet
6. 100 feet

Page 99
1. 12   12   18   5
2. $5\frac{1}{2}$ 8   3   20
3. $28\frac{1}{2}$ 18   5   7
4. 6   14   3   2
5. 4   3   3   20
6. 3 ft
7. 4 ft

Page 100
1. 5^6 4^3 y^2 x^2
2. 2^4 a^3 q^4 6^4
3. 9^3 r^3 p^1 2^2
4. s^5 1^4 15^2 t^3
5. 3^3 w^2 5^3 7^4
6. a^2 9^4 z^4 q^1
7. 4^2 3^3 x^1 4^4

Page 101
1. 16   9   4   25
2. 64   0   100   81
3. 1   1,000   32   256
4. 1,331   10   441   729
5. 196   8,000   289   0
6. 72   400   432   512
7. 12,500   2,187   12,348   25
8. 729   256   15,625   46,656
9. 1,296   0   81   1,024

Page 102
1. 11   7   12   5
2. 15   9   4   18
3. a a^2 x^3 y^4
4. ab x^2y^2 xy^2 abc
5. $3a$ $4b^2$ $6ab$ $8xy^2$
6. $13x^5y^2$ $17x^6$ $18b^2c^3$ $20ab^3$

Page 103
1. Oliver worked 8.5 hours on Monday, 6 hours on Tuesday, and 7.5 hours on Wednesday. How much longer did he work on Monday than Tuesday? $2\frac{1}{2}$ hours
2. Winona bought a car that cost $5,480, including a $199 service warranty. She paid $1,000 as a down payment and took out a loan for the rest. What was the amount of her loan? $4,480
3. Ernest works part-time and earns $75 per week. If he saves about $25 per week, how much money will he save in 6 weeks? $150
4. Iris spent $4.32 for a calculator, $12.89 for a backpack, and $8.99 for a lantern. How much more did the lantern cost than the calculator? $4.67

Page 104
1. 45
2. 5
3. 6 in.
4. $2\frac{1}{2}$ in.
5. 4 times
6. 8, 50
7. $57.50
8. 1,250 ft
9. 12 ft

Page 105

1. 13   21   3   5
2. 8   4   16   9
3. 7 $3\frac{3}{7}$ 5   18
4. 27   6   27   2
5. 100   8   256   625
6. 21 in.   7. length = 400 in., width = 300 in.
8. A (1, 1)   E (−7, 0)
    B (−2, −3)   F (−3, 3)
    C (6, −1)   G (4, −8)
    D (0, 7)

Unit 4

Page 106

1. 17%   10%   25%   62.5%   37.5%   150%   1%
                        or   or
                     $62\frac{1}{2}$% $37\frac{1}{2}$%
2. 0.15   0.001   0.25   0.20   0.125   0.01   1.25
3. 0.05, $\frac{5}{100} = \frac{1}{20}$ 0.47, $\frac{47}{100}$
4. 5   90   -15%
5. $4.50   6. $13.50   7. $130.36
8. $57   9. $2,311.04   10. $15.20
11. $932.50   12. $2,120   13. 60%

Page 107

1. $16 $19.50 $17.50
2. $25 $4.88 $43.88
3. $28.13 $6.75 $6.65
4. $12.50   5. $104.04
6. $1,273.08   7. $945
8. $140   9. $200

Page 108

1. 144   19   9
2. 28,200   78   590
3. 30   52   20
4. 600   17,800   940
5. 156   14,400   300
6. 14   30   441
7. $8\frac{1}{6}$ $6\frac{1}{4}$ $1\frac{1}{2}$
8. 7.4   0.0193   1.32
9. $2\frac{1}{2}$ $2\frac{2}{3}$ 9
10. 7.1   0.0135   0.9
11. 6   3   9
12. 36   2 $6\frac{1}{2}$
13. 300 ft   14. 260 grams
15. 5 lb 1 oz   16. 26 packages
17. 100,000 centimeters   18. 92.3 meters
19. 900 centimeters   20. $0.23

Page 109

1. $1\frac{1}{2}$ gallons   2. $1,600
3. 3,500 square feet   4. $285
5. 128 cubic ft   6. 384 bushels
7. $69\frac{1}{7}$ ft   8. $125\frac{5}{7}$ square feet
9. 6,160 or 6,157.54   10. 15 meters
   cubic feet

Page 110

1. $250   5. $14\frac{2}{7}$%
2 $10.26
3. $10,400   6. $12\frac{1}{2}$%
4. $3,000   7. 50%
                 8. 20%

Page 111

1. 3   −12   −2   −11
2. −48   88   −25   −12
3. 9   1,000   5   6
4. 9k   9m   −3ab   7n
5. 3   6   2
6. 15   2   3
7. 60 ft
8. 5 ft
9. 12 ft
10. 18
11. 12 in., 16 in.
12. 70 degrees
13. A (−3,3)
    B (4,1)
    C (1, −3)
    D (−2, −2)

14.

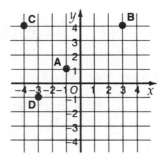

Page 112

1. $\frac{10}{40} = \frac{1}{4}$
2. $\frac{10}{50} = \frac{1}{5}$
3. $100, $150, $200   4. 45°, 60°, 75°
5. $60, $120, $180   6. $6\frac{2}{3}$ feet
7. $\frac{9}{14}$ 8. $13\frac{1}{3}$ miles
9. 14   6   3   2
10. 20   3   4   3
11. 7 $5\frac{1}{3}$ 18 $28\frac{1}{2}$

MASTERY TEST

Page 114

1. 50%   12. $116.40   17. 9,000 miles
2. 75%   13. 224,185,250   18. $37.50
3. 37.5% or $37\frac{1}{2}$% 14. $400   19. $18
4. 0.0025   15. $9,630   20. $23,000
5. 0.60 = 0.6   16. $110.25
6. 0.875
7. $\frac{3}{400}$
8. $\frac{2}{5}$
9. $\frac{5}{8}$
10. 42
11. 330

Page 115

21. 5,280
22. 64
23. 17,000
24. 27
25. 51
26. 5,700
27. 3,000
28. 24
29. 25
30. 63
31. 180
32. 120

33. $3\frac{3}{4}$
34. $1\frac{3}{4}$
35. 4.139
36. 22
37. $1\frac{3}{16}$
38. 8.06
39. $1\frac{13}{16}$
40. $21\frac{1}{4}$
41. 0.152
42. 5
43. 4
44. 3

45. 10 lb 5 oz
46. 2 gal 2 qt
47. 2 min 42 sec
48. 18 yd 1 ft

Page 116

49. 1,400 ft
50. 3,000 sq m
51. 6,000 cu in.
52. 44 ft
53. 154 sq m
54. 1,540 cu cm

55. $40
56. $w = \frac{A}{L}$
57. 180 meters
58. 490 ft
59. 13,500 sq ft

60. $108
61. 4 ft
62. 300 cubic yards
63. 30 tons
64. 9,240 cubic ft

Page 117

65. 3
66. 4
67. 4
68. 22
69. 14
70. 6
71. 10
72. 9
73. 6

74. 5
75. 13
76. 3
77. Yoko has $275, Teresa has $250.
78. Father is 45 years old, I am 20 years old.

79. length: 20 m, width: 10 m
80. 25, 50, 100
81. A(−4, 2)
82. B(4, 1)
83. C(1, −2)
84. D(−2, −2)

85. A
86. B
87. C
88. D

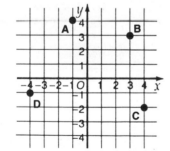

Page 118

89. 45
90. 10
91. 5
92. 5

93. 9
94. 20
95. 3:2 or $\frac{3}{2}$
96. 18 m, 27 m, 36 m

97. Diva has $12, Clara has $15, Lucia has $18.
98. 10 gallons
99. $300
100. 60 feet